The incredible Darjeeling 'B' Class

A historical and continuing story

by David Churchill

802 at Darjeeling Station on 11 March 2015 *(Colin Davies)*

Riding the cab roof of 'B' Class 780 on Loop No 2 on 2 February 1993 *(Laurie Marshall)*

Introduction and background

The Darjeeling 'B' Class is amongst the world's iconic steam locomotives – quite surprising considering they work on what the Raj termed a 'Class 2' narrow-gauge railway in the foothills of the Himalayas, that only one of them has ever worked outside India, and that they never exceeded a speed of 20mph. They have appeared over the years in countless railway and travel books and magazines, in films and on TV, on stamps and even on cigarette cards. The enduring image is of a small and rather eccentric train dwarfed by the Himalayas, with men riding on the front dropping sand and a coal breaker riding on top.

Although long affectionately termed the 'Toy Train', it is very far removed from being a toy. In 1879 the Darjeeling Himalayan Railway (DHR) was a pioneering project to economically build and work a railway in a situation that pushed existing technology to, and well beyond, proven limits. Working the line required a very special locomotive and within a few years the original designs had evolved into the 'B' Class. It is a fine example of an engineering design which developed very quickly to meet a very specific and demanding duty, and was then very difficult to better, despite repeated attempts to replace it. These locomotives, little changed, have remarkably continued to do what they were built for in commercial service for well over 100 years.

My first visit to Darjeeling was in 1979. At the time our primary reference was Brian Reed's *Profile* booklet. Although the text was excellent and very thorough, the limited pictures didn't really prepare one for the experience of travelling to Darjeeling behind a 'B' Class. I made more visits to India and kept a strong interest in their narrow gauge. After joining the Darjeeling Himalayan Railway Society (DHRS) the spark which turned interest into obsession with the DHR and its history was meeting the late Terry Martin, then working on his book *Halfway to Heaven*. I was able to offer him some help and also with the two volumes of *Iron Sherpa* that followed – the second published in 2010, shortly after his untimely death. Comprehensive and detailed as Terry's books are, there was scope for something more. As a modeller I would have liked to include more pictures of the locomotive details and variations but there just wasn't space, hence the genesis of the present book. Originally conceived as a pictorial supplement to Iron Sherpa, but one that finally grew into a stand-alone book, which the DHRS, believe to be of historical importance.

I have not attempted to cover the history and geography of the DHR in any detail, only insofar as it directly affects the 'B' Class story. Other works listed in the bibliography provide this information. A readily available and economical volume is *Darjeeling's Wonderful Railway – A Visitor's Guide* by David Charlesworth, which was published by the DHRS in November 2016. For a complete history of No 19, the only 'B' Class to leave India, one should purchase *The Story of 19B* also by David Charlesworth and published by the DHRS.

An enormous number of people have contributed to this book. I give thanks to them all.

Many offered the use of their pictures or material from their collections. I hope all are mentioned and properly acknowledged in the credits. However I would like to give a special vote of thanks to Nick Fitzgerald and Peter Jordan who have provided a great many photographs, and also to Chris Pietruski, Mike Smith of Bournemouth Railway Club, Laurie Marshall, Michael Bishop and to Werner Brutzer for the cover picture.

Malcolm Brown has always been quick with expert advice and suggestions to keep me on the straight and narrow.

G R M Raju, David Mead, Cedric Lodge, Roger Waller and Leandro Guidini kindly provided invaluable material. John Milner and Fabien Raymondaud as copyright holders have allowed me to use some material from Terry Martin's books. Peter Tiller, David Barrie, Jim Hay (NGRS), Mike Weedon, Simon Darvill, Subhabrata Chattopadhyay, Julien Webb and Kathryn Marsden, plus many other members of the Darjeeling Himalayan Railway Society and of IRFCA, assisted in all sorts of ways, perhaps without realising just how useful their contributions have been.

Staff of libraries and archives have been ever helpful – those at the British Library Asian and African Studies in particular, but also the Gladstone Library at Hawarden, The National Railway Museum Search Engine at York, the Mitchell Library, the Library of Birmingham and, via the Internet, the De Golyer Library and the Railroad Museum of Pennsylvania.

David Charlesworth has been a support throughout and has turned my motley assortment of Word docs, photographs, and drawings into a proper book, while putting up with my lateness, constant changes and revisions.

Final proof reading was by Peter Jordan, who gave up a considerable number of hours in the final stage, and Janine Bird, though the first drafts were read by Graham Bell of Click Click Words and Pictures and Jenni Croft, both experienced in writing and teaching.

And finally my wife, Diane and family who have had to live with me spending endless hours tapping away on the computer or disappearing to the British Library.

Without each of them the result would have been much the poorer. However, any errors and omissions are down to me.

David Churchill, May 2018
Solihull UK

The challenge of building and running the Darjeeling Himalayan Railway

The Darjeeling Steam Tramway (soon renamed The Darjeeling Himalayan Railway) was promoted by Franklin Prestage, who was also Agent of the East Bengal Railway, and financed on the basis of minimal capital expenditure by running along the Hill Cart Road between Siliguri and Darjeeling. The ruling gradient was thought to be 1 in 25. It was soon realised that in places the gradients were far more severe, but despite this the line was built as planned and fully opened by 1881. The problems and solutions in these early years were set out succinctly by Mr Stevenson, the DHR company secretary, who

Z reverses were introduced where it was advantageous to gain height in a short distance. Here, 797 is negotiating Reverse No 4 above Tindharia. The train is running uphill and is being flagged into the lower headshunt. It will then reverse across the road into the upper headshunt before resuming forward travel on the track on the right *(Laurie Marshall)*

wrote on 30 July 1883 to plead for amended contract terms with the Government. Some extracts are reproduced below.

….it was *believed that the gradient on the Cart-roads in no case exceeded 1 in 25 and averaged 1 in 30 and upon this assumption it was estimated that …. the works would be completed and in working order within two years. The detail surveys, which could not be undertaken until the Company was formed and funds were available, showed that instead of the ruling gradient being only 1 in 25, they were actually in many places 1 in 19. It was, however, thought best that the Company, in order to comply with its engagements with Government as to times of completion, &c, and to reduce to the utmost the serious inconvenience to road traffic during the construction of the Railway, should, in the first instance, lay its rails on the unexpectedly heavy gradients of the road, and it did so. Results in working and the traffic to be dealt with, however, very soon showed that the gradients ought to be reduced to fully what they were originally thought to be, and the amount of traffic and due regard for the public safety required that the numerous curves on the line should be made of larger radius, and that the alignment of the railway should be improved generally. The engine-power and rolling stock of the Railway, together with its numerous equipments, had to be much increased…., the Company has made very great alterations and improvements in the alignment of the Railway by the construction of reverses and loops and other works. The greater part of these are now completed,…. The Company has strained its resources to the utmost, and has spared no money in order to make their railway and its special and costly equipments as complete and perfect as possible. The difficulties they have encountered and overcome have been novel and in some respects such as could not have been foreseen.*

Another method used to moderate the over-severe gradients was to introduce the famous loops. The picture from about 1890 shows a down train at Agony Point. The loops are best illustrated by early pictures such as this, because trees and undergrowth make more recent pictures difficult. Some of the other loops had a much more complicated layout. Space restrictions meant that the curves were, and still are, extremely tight. This loop gained 22 feet in height *(DHRS Archive)*

The improvements made during the first few years involved much civil engineering with considerable re-aligning of the trackbed to avoid the most severe curves and gradients. Where the topography made this difficult or unduly expensive, the Z reverses or loops referred to by Mr Stevenson were used.

It was clear from the beginning that working the DHR would stretch proven 1870s technology to, and indeed well beyond, its limits. A comparison with the Festiniog Railway, much publicised at the time as a model for a narrow-gauge line is illuminating. The figures for the DHR apply to the 1890s after the various 'improvements'. (*The Festiniog Railway* by J I C Boyd, Vol 1 First Edition, page 10).

	DHR	FESTINIOG
Length of line	51 miles	13 miles
Expected traffic	Primarily freight, majority 'uphill' Passengers secondary	Primarily freight, majority 'downhill' Passengers secondary
Maximum gradient	1 in 25 and 1 in 23	1 in 68
Average gradient	1 in 30.72 for 40 miles between Sukna and Ghoom	1 in 92 for 12¼ miles
Minimum radius curve	60 ft radius	116 ft radius
% of line comprising curves tighter than 100 ft radius	13.4 % (that is 6.8 miles of the 51)	None
Character of line	Majority roadside or 'on road'	Almost all reserved track

Not a perfectly sharp picture, but a remarkable effort considering that it is an amateur shot from about 1910 and taken from the rear of a moving train. It emphasises the roadside nature of much of the line and it is interesting to see the condition of parts of the main Hill Cart Road at that time. The line has always had a large number of road crossings – 117 in year 2000, reduced from 383 in 1886.

The 1896 Gradient profile emphasises the severity of the DHR. It started with 7.25 miles flat(ish) from Siliguri to Sukna, then almost 40 miles climbing steadily at an average of 1 in 30.72 to the summit at Ghum, followed by about 4 miles down into Darjeeling at an average of 1 in 31.6. Note spellings are shown as at 1896 – they vary. Frequent realignments following landslips, the building of Batasia Loop and extension from Siliguri to New Jalpaiguri have changed figures somewhat since but they remain essentially similar.

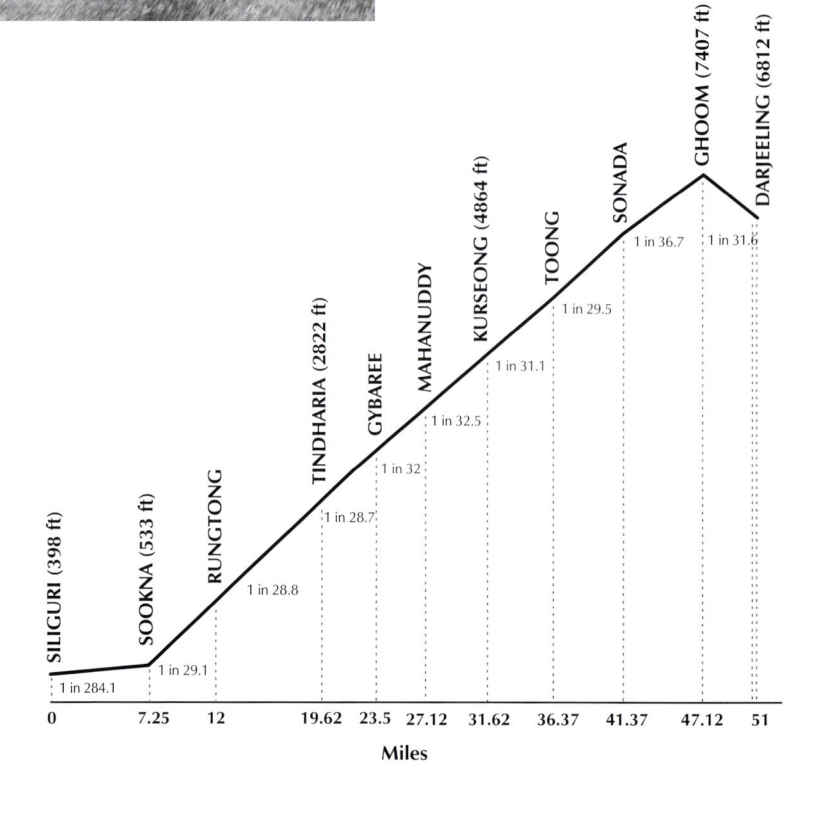

In 1979, 780 runs along the main street of Kurseong, one of the well known 'bazaar' sections of the line. These have always been difficult sections to work, made worse by modern traffic.

Also in 1979, 790 on an up train waits in one of the passing sidings while 780 on a down train runs through. Passing sidings rather than passing loops were used on the hill section to reduce the civil engineering work needed, except at the main intermediate stations of Kurseong, Tung and Ghum. Among other things, the length of the passing sidings restricted the train length that could be worked *(Chris Pietruski)*

Evolution of the 'B' Class – what came before

Schedule B of the contract to allow building of the Darjeeling Steam Tramway signed in April 1879 specified that *'The Company is to provide in the first instance…. eight locomotive engines capable of hauling 9 tons each up the ruling gradients of the cart-road at 7 miles an hour….'*. Franklin Prestage's original intention seems to have been to operate the line with back-to-back pairs of locomotives, rather optimistically with one driver. It is not known which manufacturers were approached to supply the locomotives. There is an intriguing report in the *Western Daily Press* of 25 March 1879, a few days before the contract was signed, where a visitor to Darjeeling states he attended a meeting held for the purpose of Mr Prestage giving an account of the scheme for the tramway. At that meeting were shown *'drawings and specifications for the locomotives from Messrs Fox, Walker and Company, Bristol'*. Nothing further has yet been uncovered as to what locomotive design these were for.

It interesting to speculate, in view of Prestage's known links with Portmadoc, whether Fairlie locomotives, much publicised and successful in North Wales, might have been considered. There seemed to be resistance to them in India – the first there being ordered in 1879 to work a proposed military metre-gauge line which had a chequered but mostly inactive career. Probably the curves of the DHR would have made a Fairlie impracticable anyway.

By May 1879 Sharp Stewart of Manchester were the favoured supplier mentioned in a report in *The Englishman*, although the expectation still seemed to be that back-to-back pairs of locomotives would be used. That the order went to Sharp Stewart is perhaps not surprising: they had supplied a fair number of broad and metre-gauge locomotives to India in the 1860s and 70s, including all the early locomotives for the Eastern Bengal Railway, where some of those with links to the DHR had been or were still employed. They also had some experience of narrow-gauge locomotives.

The 'No1' Class – later known as 'C' Class

Using back-to-back pairs of locos seems to have been quietly forgotten and Sharp Stewart produced individual locomotives – the DHR's 'No1' class. These were a small side-tank locomotive with outside frames stepped outward aft of the driving wheels to enable a wider firebox and 8in x 14in cylinders. Eight were delivered in 1879/80. The first four may have been supplied with 7in x 10in cylinders – records are inconclusive but later DHR reports suggest they all had 8in diameter cylinders.

These proved to be barely capable of working the line, although the builder can hardly be blamed due to the much more severe grades and sharper curves than were specified.

Close up pictures of the No1 class in service on the DHR are very rare. It is interesting that the loco carries no indication of ownership or line number, presumably the works plates were sufficient identification with only eight locos. There are some modifications apparent from its state as delivered – whistle on dome, sandboxes removed, re-railing beam slung along tank side, plate on left back of footplate to reduce coal spillage.

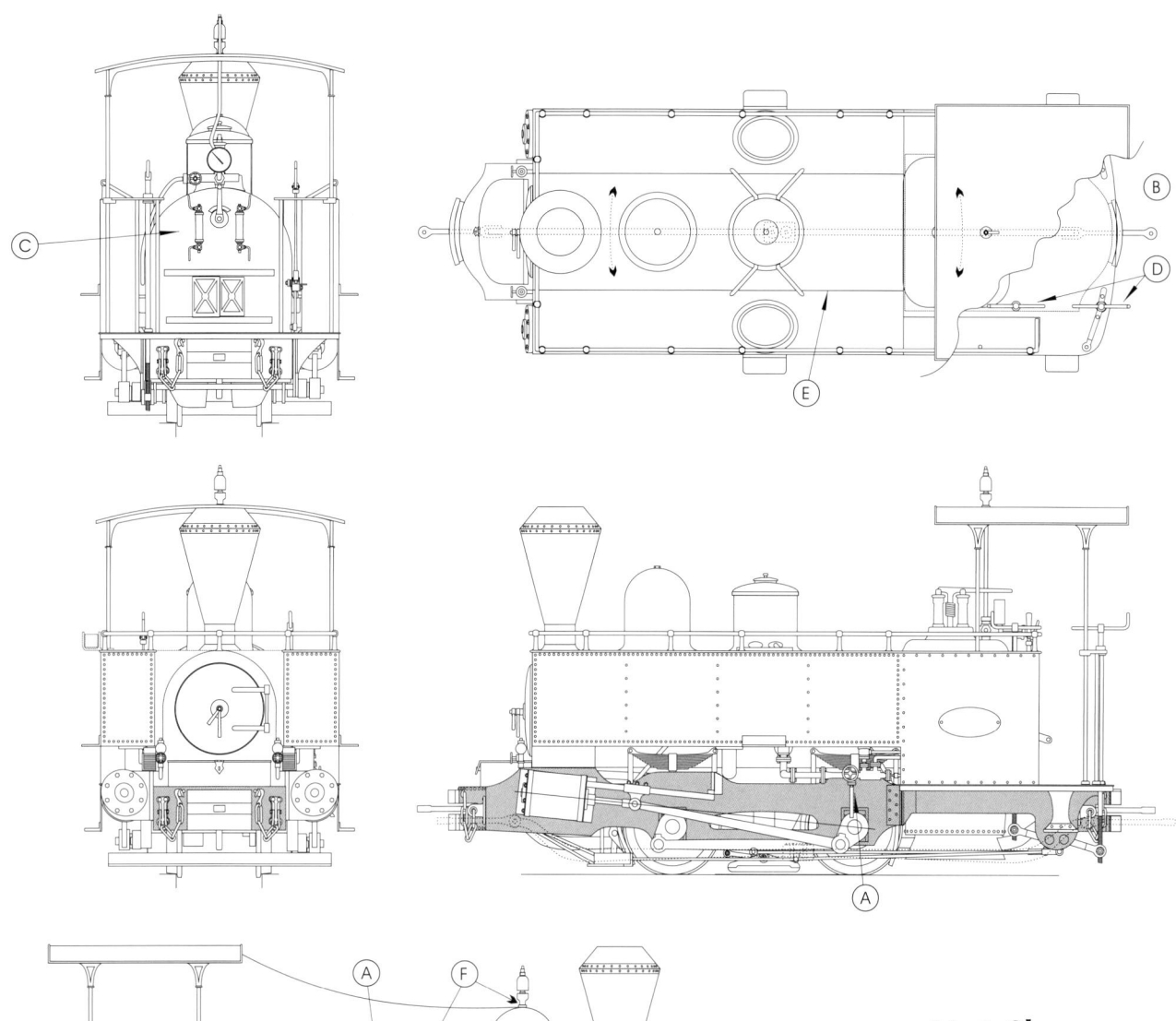

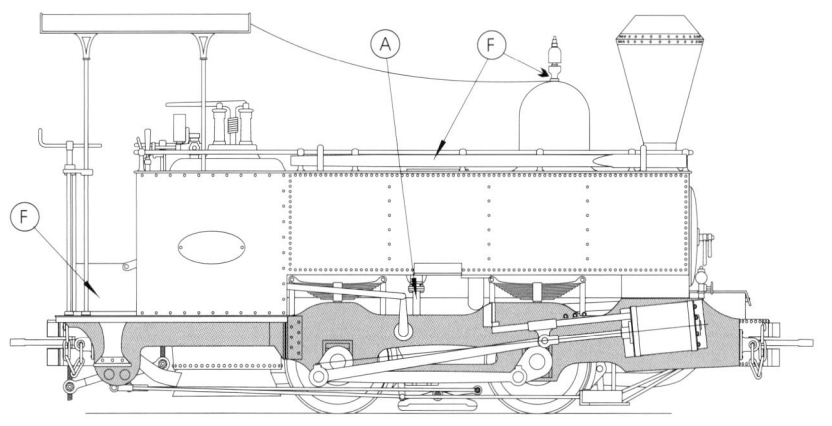

No1 Class

The left-hand side elevation shows No 7 as built and is based on the maker's g/a drawing and photographs. The right-hand side elevation shows modifications on a loco in service during the mid-1880s. Note the unusual centrally pivoted draw-bar shown dotted on the left-hand side elevation and plan.

(A) Injector on left side only. A feed-pump driven from an eccentric on the rear axle was on the right side. The pipework/controls to this are uncertain.

(B) Roof cutaway to show cab and (dotted) rear buffer profile.

(C) Backhead details uncertain and drawing part based on later locos.

(D) The front brake handle operated the skid brake, the rear the conventional brake. The controls for the Le Chatelier counter pressure brake are uncertain.

(E) Unsure to what extent the boiler/tank tops were plated over, although they are in the surviving Brazilian locomotive.

(F) Modifications while in service included:-

Removal of sandbox, probably relying on manual sanding when required.

Whistle moved from cab roof to dome.

Re-railing beam slung along right-hand side (shown dotted on front elevation).

Addition of plate at left-hand side cab rear to prevent coal spillage, and a raised wooden floor to the cab.

0 1 2 3 4 5 6 7 8 9 Feet (1:48)

©David Churchill 2018

The No1 Class locos had a fairly short life on the DHR but Sharp Stewart built two further locomotives to the same drawings with works nos 3486 and 3487 of 1888. The only change noted in the order book was omission of the skid brake. The order was from Brazilian agents Fonseca & Machado and the locomotives worked in the Sao Paolo area for the Companhia Paulista de Estradas de Ferro and were finally numbered 950 and 951.

Although soon superseded by the No2 class, some No1 Class were kept until the 1890s to work over a weak bridge near Siliguri. All were eventually sold on to the North Western Railway, and seven of the eight finally ended up at that railway's Dandot Colliery. This unidentified loco is at Dandot in 1907, little changed from when it worked on the DHR years earlier *(H V O Waters – collection Neville Thomas)*

Remarkably 951 still survives in Brazil in the Meseu Eduardo Andre Matarazzo, Bedebouro, Sao Paolo State. It is in poor condition but not greatly changed from its original state, [left] as in 1968, [above] more recent views *(Leandro Guidini)*

The 'No2' Class – later known as 'A' Class

After the experience with the 'No1' Class a rapid rethink must have followed and Sharp Stewart soon came up with a radically changed design – The 'No2' (later the 'A') Class. Eight were delivered in 1881/82. They all had Sharp Stewart works plates and works numbers, although four were sub-contracted to Hunslet of Leeds. The sharp curves still required a four-wheeled design with a maximum wheelbase of 4ft 3in and maximum axle load was restricted to 6½ tons.

Major changes were;

1. Use of a well tank situated between the frames to lower the centre of gravity.
2. This necessitated the use of outside valve gear – Walschaerts was chosen, which was unusual in the UK at that time.
3. 10in x 14in cylinders.

The result was a more powerful locomotive which was better balanced for the narrow-gauge and less than immaculate track. The new locos proved successful; an 1885 inspection report noted that *'At first, derailments of locomotives were frequent (referring to the earlier 'No1' class) but by distributing the weight more evenly on the frame and by placing tanks below it, thus keeping the centre of gravity low, the danger has almost disappeared.'*

Further modifications were soon made locally, perhaps at Tindharia but more probably at the East Bengal Railway works at Kanchapara. In Col. de Bourbel's report of an inspection in April 1885 he recorded that *'Seven engines [of the eight] have now been fitted with collar tanks and coal bunkers. The former holds 80 gallons, in addition to the well tank, which holds 250 gallons. The coal bunkers carry 18 maunds of coal. The total load of engine is now 12 tons 10 cwt on four wheels with 4 inch axle and 4 inch journal.'* The term 'collar tank' was used for what is now usually termed a saddle tank. As well as providing much-needed extra fuel and water capacity, the tank and bunker helped to further improve the 'balance' of the loco. With these additions the locomotives were beginning to look more like the familiar DHR 'B' Class still in use 130 years later.

Prestage made an interesting comment in the *Railway Engineer* when he referred to *'Dr. Anderson of the Baldwin Locomotive Works, Philadelphia'* being *'a very good friend to the undertaking '*. It is not clear who Dr Anderson was or what he did, although it may be connected with the use of a very wide firebox compared with the track gauge.

The 'No2' Class locomotives sub-contracted to Hunslet were built alongside CHARLES – the first of the Penrhyn 'Ladies'. It has been suggested that there may have been a connection with the design of the 'B' Class. This seems unlikely as, apart from some similar dimensions and looking superficially similar, they could hardly be more different. The 'Ladies' have Stephenson's valve gear, no well tanks, no widening of the frames to accommodate a wider firebox, and their boilers are classic Hunslet design. The evolution of the Darjeeling

'B' Class through the 'No1' and 'A' class seems a quite clear and logical response to the need for a more powerful, better balanced design to meet traffic demands and take advantage of the improvement of the track. Perhaps seeing CHARLES might have given someone (the DHR's consulting engineer?) the idea of adding the saddle tank to the 'A 'class as this was done quite soon after they arrived in India

An enlargement from a very early (1885) photograph of No2 Class No 15, in service. The engine has already been modified with the collar tank and bunker. No 15 was built by Sharp Stewart in 1883, the penultimate example of the class. The last six of the eight locos had a full-width spectacle plate *(Collection Simon Creet)*

No 10 was the second 'A' Class built, here seen in a picture thought to be at Siliguri shed and dated as circa 1902. The first two locos did not have the full width cab front sheet. The crosshead-driven feed pump and its associated pipework is visible. An interesting detail is that the re-railing beam slopes forward slightly *(Collection Roger West)*

'No 2' – the 'A' Class

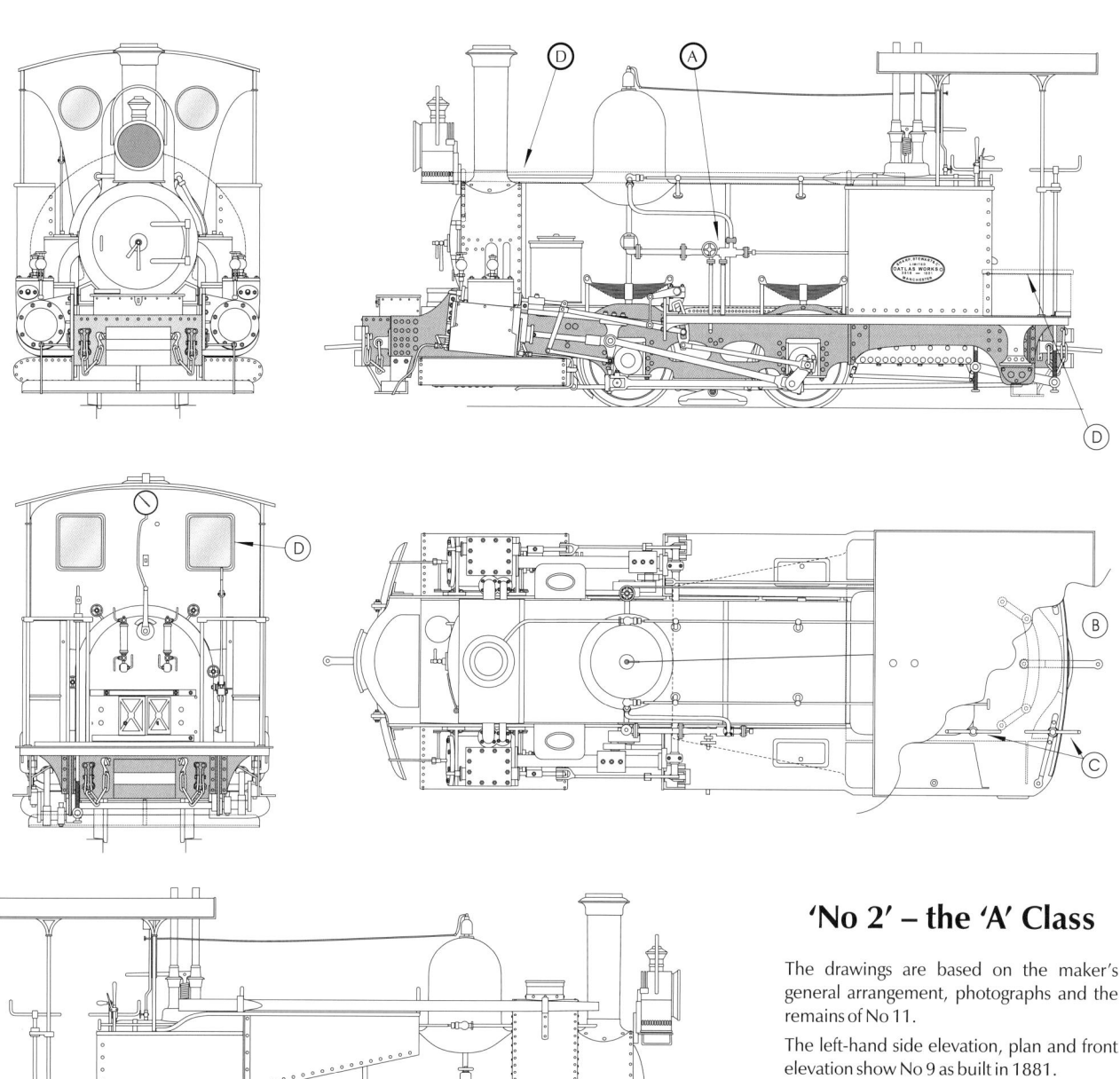

The drawings are based on the maker's general arrangement, photographs and the remains of No 11.

The left-hand side elevation, plan and front elevation show No 9 as built in 1881.

The rear elevation shows the full width cab front as fitted to Nos 11-16.

The right-hand side elevation illustrates a rebuilt locomotive with added saddle-tank and coal bunker as in service c.1890.

The approximate profile of the tank is shown dotted on the front elevation and that of the bunker is dotted on the plan.

Note that the route of the pipes connecting the saddle to the well-tank is not clear. They are not visible on photographs and did not enter the well tank forward of the smokebox as on the 'B' Class.

(A) The Injector was on the left-hand side only with a crosshead driven feed-pump on the right side. The pipework and controls to this are somewhat uncertain.

(B) Roof is cutaway to show cab, rear buffer profile and the coupling arrangement.

(C) The front brake handle operated the skid brake, the rear the conventional brake. The controls for the Le Chatelier counter-pressure brake are uncertain.

(D) Later-built locos had a full-width cab front sheet with rectangular windows as shown on the rear elevation. Other changes included the addition of a plate at the left-hand side of the cab rear to prevent coal spillage, and a re-railing pole.

0 1 2 3 4 5 6 7 8 9 Feet (1:48)

© David Churchill/David Charlesworth 2009

13

'A' Class No 11 pictured in an excellent amateur photograph taken in 1907 at Siliguri. By that time the 'A' Class were being withdrawn, with only three still in service.

By 1911, only two were left on the DHR although these continued in use for many years. The loco has a plain top to the chimney rather than the usual cap, which was sometimes seen on 'B' Class at that time *(Collection Guy Comer)*

The last 'A' Class withdrawn was No 9 in 1954. It had been updated during its long life and in this picture from the late 1920s or 30s, it carries a 'rebuilt Tindharia 1914' plate. The updates were similar to those on the early 'B' Class including safety valves on the dome, a whistle on the firebox and a side raking ashpan. There are no balanced outside cranks as yet but it does have the bunker changed to the rectangular 'B' Class style *(Bournemouth Railway Club Trust, Kelland collection 50082)*

Although No 9 was withdrawn in 1954, the remains of another one still exist! These are believed to be of No 11, which certainly survived into the 1940s. They were dumped in the former sidings at Siliguri Junction and when it was photographed in 1979 the rear frames, footplate and tapered style of bunker, which bore the number 11A, were still in place.

Unfortunatley, it was not until the formation of the Darjeeling Himalayan Railway Society in 1997 that the importance of these remains was realised.

The remains were finally rescued from Siliguri Junction in 2004 and taken to Tindharia for safe keeping. Much had been lost since 1979, but what has survived of 11A is still of enormous historical interest. The picture [below] shows the well tank between the frames (the top and bottom plates of tank have been lost) with wider section at front and narrower at the rear to clear the wheels).

The cylinder motion bracket and both views show surviving back of the wing tank. Balanced outside cranks have been fitted.

The rear section of the frames is lost but the fabricated supports that carried them 6 inches wider apart than the front section can still be seen. Although it is an 'A' Class, this illustrates some otherwise lost features of the early examples of the 'B' Class) *(Pictures by the late Terry Martin, courtesy Fabien Raymondaud, and David Charlesworth*

From 'No2' Class to 'No3' (later 'B' Class)

The 'No2' Class locomotives proved very satisfactory, but within a few years booming traffic figures (both passengers and goods carried had increased by over 50% between 1882 and 1888) required a still more powerful locomotive.

Prestage wrote: *'I would suggest lengthening the boiler and giving bigger cylinders in proportion. The engines would be better balanced particularly for a line like this with a steady ascent all in one direction.'* For goods traffic he proposed *'also the same class of engine but two coupled together in the manner proposed by Mr Garrard and forming what that gentleman terms "a climber".* G M Garrard was the company's inspecting engineer in England (and related by marriage to Prestage). 'The Climber' was another variation on the double locomotive and was described and illustrated in the technical press of the time.

Perhaps, fortunately, the DHR management opted solely for Prestage's first option: the development of the 'No2' class into what became Sharp Stewart's 'No3' or 'B' Class. By now, with improved track and alignment of the line, the introduction of loops and reverses and operating experience allowed a longer wheelbase (5ft 6in) and a heavier locomotive. This permitted what was essentially an enlarged version of the 'No2' class.

Cylinders were now 11in x 14in and water and fuel capacity increased. At first called the 'No3' Class, within a few months (by September 1889) they were being referred to as the 'B' Class and the name and the distinctive 0-4-0 tank design with large overhang at front and rear is still in use today.

Summaries of the 'No1', 'No2' and the first batch of 'No3' Classes of locomotive

	NO 1 (C) CLASS	NO 2 (A) CLASS	NO 3 (B) CLASS
Wheelbase (ins)	51	51	66
Wheel dia.(ins)	26	26	26
Cylinder bore x stroke (ins)	8 x 14	10 x 14	11 x 14
Boiler pressure (Psi)	140	140	140
Tube length (ins)	90 55 tubes 1⅝ in dia.	96 65 tubes 1⅝ in dia.	117
Boiler dia. (ins)	30.5	30.75	33
Boiler centre line from rail (ins)	44	46	46
Grate area (sq. ft.)	5.4	7.3	8.8
Heating surface firebox (sq. ft.) Boiler (sq. ft).	28 175	36 228	40.4 276
Water capacity (galls)	250	250 well / 75 saddle 20 wing	260 well / 120 saddle 20 wing
Weight in working order (tons)	10.1	12	13.35
Max axle load	5 tons 10 cwt	6 tons 10 cwt	7 tons 10 cwt
Tractive effort at 75% BP (lbf)	3,618	5,653	6,841

A comparison between the performance of the 'A' and the 'B' Classes

The information is taken from an 1897 report on 'Feeder Railways'. The vehicles are small and four-wheeled. It is interesting to see that the improved haulage capacity of the 'B' Class was achieved with little increase in coal consumption.

	'A' CLASS	'B' CLASS
Coal consumption (lb/mile) Average of up & down hill	38	39
Water consumption (gall/mile) Average of up & down hill	40	45
Usual train load – passenger	Total 13 vehicles comprising One luggage van / One 1st class carriage / Two 1st class trolleys / Four 2nd class carriages / Four 3rd class trolleys / One mail or brake van	Total 16 vehicles comprising Two luggage vans / Two 1st class carriages / Two 1st class trolleys / Four 2nd class carriages / One 2nd class trolley / Four 3rd class trolleys / One mail or brake van
Goods load uphill	5 loaded vehicles	7 loaded vehicles
Goods load downhill	12 loaded vehicles	14 loaded vehicles
Cost of engine	Rs.12,350	Rs.19,000

'B' Class locomotives – Numbers and Dates

The table lists all the 'B' Class locomotives built and includes details of builder, DHR or other railway number, Indian Railways number and the current location of those that survive.

BUILDER	WORKS NO/ DATE	DHR NO	INDIAN RAILWAY NO	TDH REBUILD PLATE [1]		CURRENT LOCATION (2018)	CURRENT BOILER NO & DATE
SS	3516/1889	17	--------	1912	Condemned [2] 1952	----------	----------
SS	3517/1889	18	777	1917	Withdrawn 1974	Delhi Museum	----------
SS	3518/1889	19	778	1908 2004 (TYS)	Withdrawn 1958. Sold to USA in 1962. Donated to Hesston in 1982. UK in 2003 [3]	Beeches Light Railway, UK	1889
SS	3519/1889	20	--------		Condemned [2] 1952	----------	----------
SS	3882/1892	21	779	1912 1922	-------------	DHR	14/L/NG 1958
SS	3883/1892	22	780	1914	-------------	DHR	01/L/NG 2007

16

BUILDER	WORKS NO/ DATE	DHR NO	IND.RLY. NO	TDH REBUILD PLATE*		CURRENT LOCATION (2018)	CURRENT BOILER NO & DATE
SS	4197/1896	23	--------		Condemned [2] 1952	---------	---------
SS	4560/1899	24	781		To Tipong 1970	Remains combined with 784 which now carries No 781.	---------
SS	4561/1899	25	782	1939	--------------	DHR	27/L/NG 2006
SS	4562/1899	26	783	1926	Withdrawn 1999	RCF, Kapurthala	---------
SS	4649/1900	KSR No 1	---------		To Dandot 1908 Scrapped	-----------	---------
SS	4650/1900	KSR No 2	---------		To Dandot 1908 Scrapped	-----------	---------
SS	4977/1903	27	784	1937	To Tipong 1970	Tinsukia now numbered 781	---------
SS	4978/1903	28	785	1942	Withdrawn 1997	Dehradun	---------
NBA	16211/1904	29	-------	1941	Condemned [2] 1952	----------	---------
NBA	16212/1904	30	786	1944	--------------	DHR	10/L/NG 1942
NBA	20143/1913	32	787	2003 (GOC)	To GOC 2002 for oil firing	DHR	05/L/NG 1928
NBA	20144/1913	33	788		--------------	DHR	6/L/NG 1954
NBA	20638/1914	34	789		To Tipong 1970	Tipong	---------
NBA	20639/1914	35	790		Withdrawn 1997	Katihar	---------
NBA	20640/1914	36	791	1963 2004	--------------	DHR	24/L/NG 1936
BLW	44912/1917	39	792		--------------	DHR	02/L/NG 2006
BLW	44913/1917	40	793		Withdrawn 1999	Chennai museum	---------
BLW	44914/1917	41	794		--------------	Matheran	
TDH	1919	42	795		--------------	DHR	03/L/NG 2007
TDH	1923	43	796		To Tipong 1970	Tipong	---------
TDH	1925	44	797		Withdrawn 1997	RDSO, Lucknow	---------
NBQ	23291/ 1925	45	798		Withdrawn 1997	Howrah museum	---------
NBQ	23292/ 1925	46	799		Withdrawn 1996	Rail Bhavan, New Delhi	---------
NBQ	23293/ 1925	47	800		Withdrawn 1997	Lucknow (Charbagh)	---------
NBQ	23300/ 1925	52 was RFT B1	805		To DHR 1943	DHR	06/L/NG 2007
NBQ	23301/ 1925	48 was RFT B2	801		To DHR 1927 Withdrawn 1990	New Bongaigaon	---------
NBQ	23302/ 1925	51 was RFT B3	804		To DHR1943	DHR	04/L/NG 2006
NBQ	23303/ 1925	53 was RFT B4	806		To DHR 1943	DHR	05/L/NG 2008
NBQ	23304/ 1925	50 was RFT B5	803		To DHR 1943 Withdrawn 1998	Moradabad	---------
NBQ	23678/ 1927	49	802		--------------	DHR	12/L/NG 1944
GOC		--------	1001 now 01		--------------	DHR	11/L/NG 2003
GOC	2004	---------	1002		Returned to GOC 2006	GOC? (dismantled?)	---------

Notes

1. Indicates locomotives which are known to have carried at some time a Tindharia rebuild plate and its date. The only plates that are still carried (2018) are on No 19, 777 and 791.
2. Locomotives "Condemned" were not always scrapped outright. They were often used for parts and thus dismantled over a period of time. In which case, one can always imagine that some part of these early withdrawals may still exist on a locomotive!
3. 19B was sold to Elliott Donnelly and shipped to Lake Forest, USA in 1962 and overhauled. Moved to Hesston in 1967, returned to Lake Forest before 1975 and donated to Hesston Steam Museum 1982. Sold to Adrian Shooter in 2003, moved to England and restored.

Abbreviations

BLW   Baldwin Locomotive Works, Philadelphia.
GOC   Golden Rock Works, Tiruchchirappalli
KSR   Kalka-Simla Railway
NBA   North British Atlas Works, Glasgow.
NBQ   North British Queens Park Works, Glasgow.
RCF   Rail Coach Factory
RDSO   Research Design and Standards Organisation
RFT   Raipur Forest Tramway
SS   Sharp Stewart, Glasgow.
TDH   Tindharia Works, Darjeeling Himalayan Railway Co.
TYS   Tyseley Locomotive Works, Birmingham, UK.

No 49 built in 1927, was the final 'B' Class, until the appearance of the Golden Rock new builds in 2004. She is pictured at Kurseong station in the 1940s.

1001 was the first of the Golden Rock 'new builds' and is seen here at Siliguri Junction shed on 15 February 2007 awaiting trials of the Ffestiniog oil firing system. *(Dave Priestley)*

The History of the 'B' Class – The first 60 years 1889-1948

The Early years

The first order (E941) for four of the 'No3' Class was placed with Sharp Stewart in 1888. They were very soon renamed as the 'B' Class. The notes recorded in Sharp Stewart's order book were as follows:

- Outside cylinders 11" x 14"
- Boiler 2'-6¼"x 9' 9" B Y Iron**
- Copper firebox heating surface 40.5 sq. ft.
- 65 Brass tubes 1⅝ "dia. HS 276 sq. ft.
- 4 Wheels coupled 2' 2" dia. cast steel
- Wheelbase 5'-6"
- Tyres & Axles Bessemer
- One No5 injector, one pump from crosshead
- Stradals drawbar
- Walschaert motion
- Saddle tank and tank between frames 400 gall
- Screw brake & skid brake
- Thickened flanges: lubrication to front flange
- Head lamp

Generally to our specification. In details generally follow E810 Delivery to be as soon as possible.*
Additional notes added later: Engine to have steam sanding gear 3 October 1888. Nos 1B, 2B, 3B, 4B 26 April 1889.

*E810 was the first order for the No2 'A' Class.
**B Y Iron is Best Yorkshire Iron.

The locomotives were delivered in 1889. Although it was specified they be numbered 1B, 2B, 3B and 4B (and 1B was photographed as such by Sharp Stewart), it seems that they were very soon renumbered as 17, 18, 19 and 20 after arrival on the DHR. Later orders for another 10 engines followed gradually over the next 15 years as listed here, with relatively few changes made to the specification (Changes recorded in Sharp Stewart order book are shown in italics):

E1010 date 15 June 1892 *Two tank engines 'B' Class as E941 but with slight modifications – see notes with letter 9/6/1892. No steam sand gear / axles strengthened to sketch / tyres special steel.* Delivered 1893.

E1090 date 1896 *One tank engine as E1010. Framing modified / For this and other remarks see notes with order / inspection Mr. Garrard.* Delivered 1897.

E1152 date 10 February 1899 *three Tank engines exactly as E1090 / Inspection Mr Garrard /see remarks in Specn.* However, the G A drawing notes change as follows, presumably made after the order but before delivery; *blast pipe top altered / piston rod increased to 11⁵⁄₁₆ in dia. / connecting and coupling rods strengthened / injector now Gresham & Craven automatic restarting No5 / skid brake not needed.* Delivered 1900.

The first 'B' Class numbered 1B, as photographed by Sharp Stewart before delivery. No evidence has been found to show that they ran in India with these numbers. This picture shows the first arrangement of blower pipe passing over the saddle tank, which was later changed.

No 28 is taking water at Tindharia in about 1910. Built in 1903, this was the final 'B' Class from Sharp Stewart before their amalgamation into North British.

E1204 date 21 May 1902 *Two four-wheeled tank engines 'B' Class / As E1152 (ie without skid brake) / For modns in front end of frame see letter and sketch recd. 27 Feb 1902 our letter 17 April 1902 / Construction of hind end of frame also to be revised.* Changes noted on the GA drawing were frame and stretchers altered, Beams ⅜ in thick. Delivered 1903.

L34 ordered 1903 (two locomotives). Sharp Stewart were now part of North British. The GA drawing does not record any further changes. Delivered 1904.

The first 'B' Class, now No 17, posed on the bridge at Agony Point, probably during the first decade of the 20th century. The crew seem to have moved slightly during the exposure. The engine is slightly updated, for example: a revised blower arrangement, Wells light on cab roof, angled supports to cab front and the plain chimney cap. In *Iron Sherpa Vol 2* Terry suggested that this engine was in black – this seems doubtful as the darker lining is clearly visible on the original print.

Not a particularly good quality picture but this view from the early 1900s is unusual as we know the name of the two gentlemen, both drivers on the DHR. Next to the engine at Woodlands is the very long serving Mr. Duncan, and the other is P J Wright. The loco appears to be No 27, which was delivered in 1903. It has the smaller works plate carried by the later Sharp Stewart locos.

A picture published in the *Indian Railway Gazette* which gives a view of the right hand side of No 28, the other 'B' Class delivered in 1903. These engines still retained the crosshead drive feed pump: the blower is now in its later position with the steam pipe passing through the right side of the saddle tank.

No 18, the second 'B' Class built in 1889, is pictured in the 1900s. There are various small changes from 'as delivered' condition, for example the plain chimney cap used on several locos at that time. The skid brake has been removed and the Wells light is just visible on the cab roof, with its feed pipe running up from the behind the right-hand side bunker.

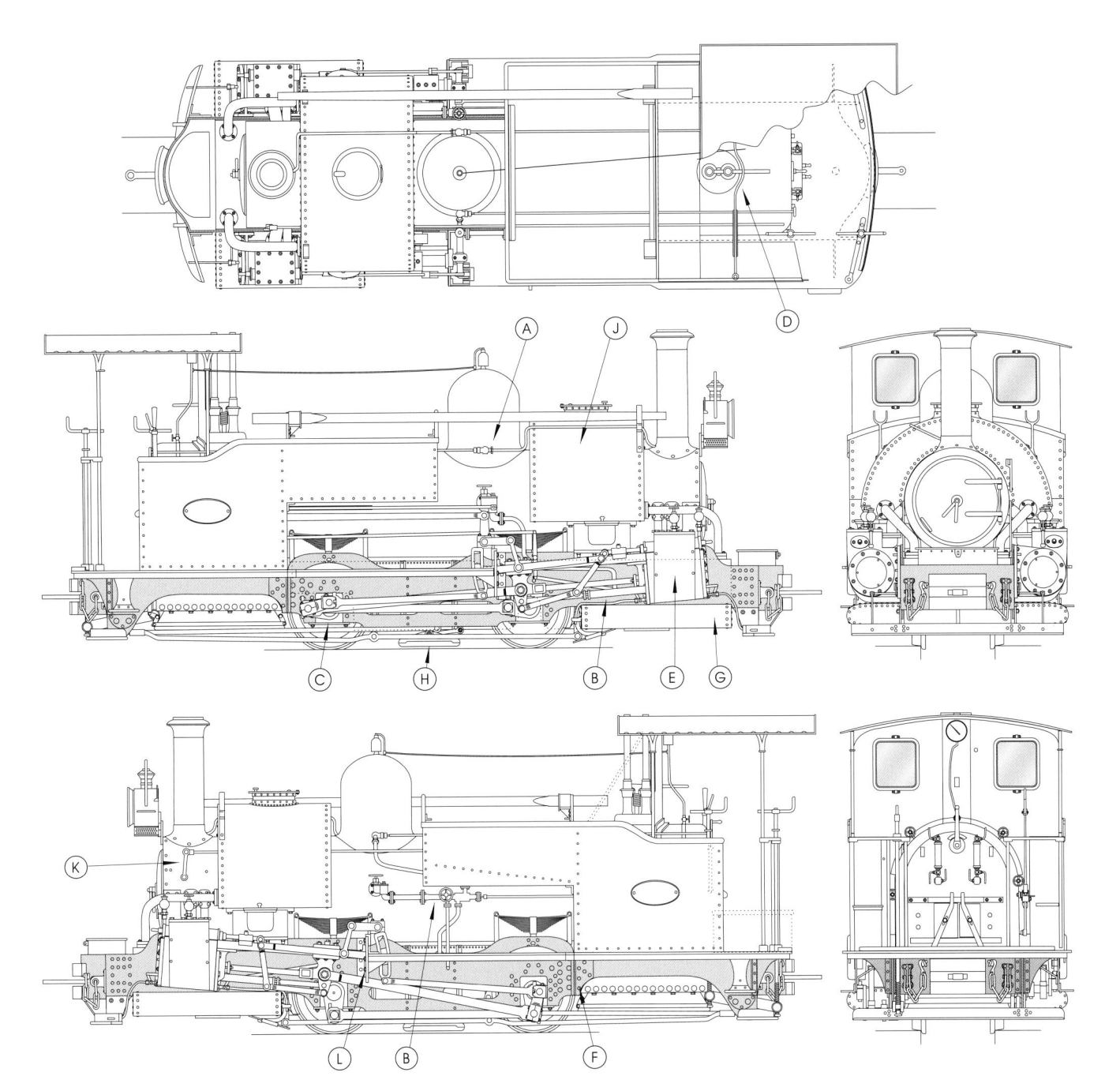

- (A) The lower valve on the side of dome and the pipe runs over the tank. This was changed to run across the boiler top and through the tube on the left hand side of the saddle tank vacated by removal of linkage K. Later the blower pipe ran through a tube through the right-hand side of the saddle tank.
- (B) Crosshead driven feed pump on the right-hand side. The injector on the left-hand side.
- (C) No balance weights on cranks.
- (D) Curve in the cab front sheet to clear the safety valve base.
- (E) Early pattern of cylinder with displacement lubricators.
- (F) A fabricated bracket where the frame widens for the firebox.
- (G) An under-cylinder 'wing' water tank.
- (H) Skid brake – removed from the mid 1890s.
- (J) A shorter saddle tank than later locomotives.
- (K) Linkage to operate blast pipe flap top – removed early.
- (L) Motion bracket fabricated from plate and angle.

'B' Class 1889

The drawing shows No 17 as built by Sharp Stewart in 1889. It is based on the makers GA drawing and works photographs and other early photographs.

Additions and changes made during the early life of the locomotives are shown dotted on the left-hand side elevation. These included extra handrails, a plate to prevent coal spillage at cab left side rear and supports for the cab front.

The dotted line on the right side elevation shows the profile of the well tank, and that on the plan shows the rear frame profile.

0 1 2 3 4 5 6 7 8 9 Feet (1:48)

©David Churchill 2018

The Kalka Simla locos

The 2ft 6in gauge Kalka-Simla Railway was finally opened in 1903 as a narrow-gauge extension of the Delhi-Umballa-Kalka Railway after much vacillation over its gauge by the authorities. Two 'B' Class locomotives were ordered from Sharp Stewart on 26 April 1899, under Order E1157 for use during construction. They were specified as being 2ft gauge and *'generally as E1152'* (which was the DHR's order of a couple of months earlier) *'but with steel boiler plates (no skid brake).'*

Given works Nos 4649 and 4650 of 1900 they were supplied to India as 2ft gauge, but converted to 2ft 6in gauge after arrival, presumably at the D-U-K Railway's works. The work was obviously carefully reported back to Glasgow and noted on the drawing as *'The main frames have been moved out from 2ft 11¼ to 3ft 5¼ and cylinders and all machinery attached to frames have been moved out with the frames. The platform has been made 6ft 6in wide. The tank under boiler along this part remains, longer gussets have been put in. No alterations whatever have been made in hinder part of engine. Wheels moved out to suit 2ft 6in gauge. New axles provided same diameter as before'.*

The KSR 'B' Class were recorded as having cost £1,233 FOB in England, rising to Rs.22,272 when erected in India. The weight in working order was quoted in a report as being 14 tons, with axle loads of 7 tons on front and rear wheels, somewhat different from that given for the DHR locos, perhaps it was an approximation.

The Kalka-Simla Railway required more powerful engines and the 'B' Class, named KALKA and SIMLA, were used for construction trains and for shunting. Working of the line was transferred to the North Western Railway in 1907 and soon afterwards the two locos were moved to the North Western Railway's Dandot Colliery in what is now Pakistan, which had previously acquired most of the ex DHR 'No1' Class. This would have required their conversion back to 2ft gauge.

The NWR closed Dandot in April 1911 but it was purchased by a local firm. However, the two 'B' Class locomotives were advertised for sale in the *Indian Railway Gazette* in June 1911. What became of them is unknown; presumably they were scrapped long ago.

The 'Revised' 'B' Class – North British 1912

From 1903, there was a break of nine years in new 'B' Class orders. By 1912, the DHR's Garratt, delivered in 1911, was clearly not going to be of much use, at least in the short term and there was also the prospect of additional motive power being needed to work the forthcoming Kishenganj and Teesta Valley branches.

The DHR ordered two further 'B' Class engines on 16 August 1912 (order L525), which were delivered in 1913 as Nos 32 and 33. For this order North British revisited the design and produced an updated version with many detail alterations, described on some drawings as 'revised 'B' Class'.

These are poor images but are the only ones found showing a Kalka Simla 'B' Class. The Kalka-Simla was a very different railway to the DHR. It was far more heavily engineered and when opened had 103 tunnels, the longest 1250 yards long. The ruling gradient was 1 in 33 but as it did not have the tight curves of the DHR it was soon able to use heavy 2-6-2 T locomotives and operate as a 'miniature mainline'.

A North British photograph of works number 20144 of 1913 which became DHR No 33. This was from the first order for the 'revised design' which had numerous detail alterations from those built previously. For details see the main text. (*Glasgow Museums & Libraries Collections, the Mitchell Library: Special Collections*)

- Original line of frames.
- Original line of running board.
- New line of running board.

Kalka-Simla gauge conversion

A Sharp Stewart drawing summarised the changes made when the Kalka-Simla locomotives were converted to 2 ft 6in gauge. The drawing illustrates the changes but some questions remain regarding the front of the locomotive.

The front sections of the frames, with the 'machinery attached to them' were moved outwards 3 inches on each side with the gussets supporting the tank lengthened to suit. To the rear of line X-X the loco was unchanged. The width of the running plate was increased by 6 inches.

No clear photographs have been found and the front view is a impression of a 2ft 6in gauge 'B' Class, assuming that the original saddle tank was retained.

©David Churchill 2018

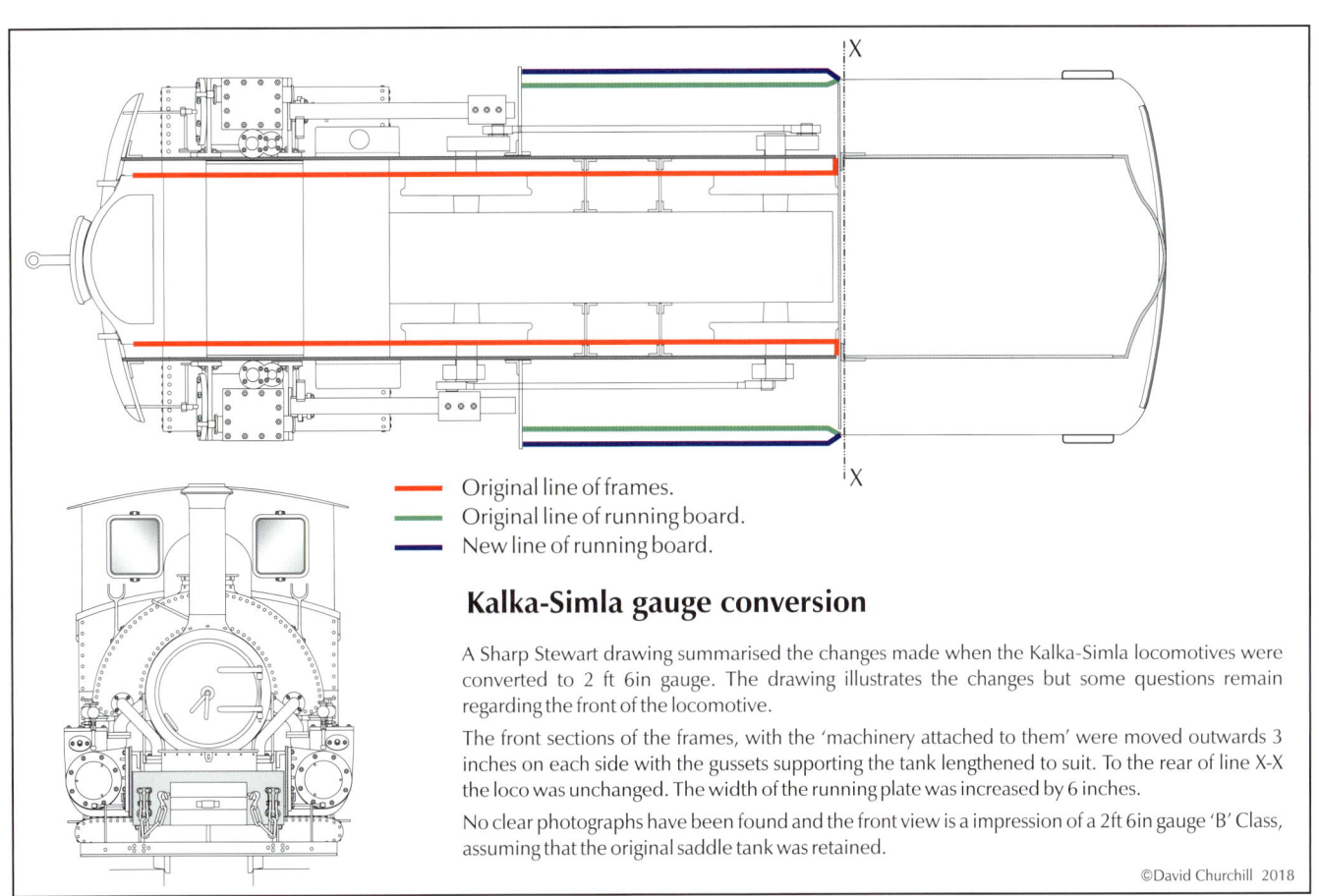

No 33 is the second example of the 'revised B', the date is unknown but is probably in the 1930s. The superb load of coal must greatly exceed the official capacity and the size of some of the lumps illustrates the need for the coal breaker. The boiler still has the older short smokebox and also has the safety valves mounted transversely on the dome. A side-raking ashpan has been fitted, also a sight feed lubricator in the cab and piston tailrods. The lamps with the loco number were normal at that time. *(Bournemouth Railway Club Trust, Kelland collection 50091)*

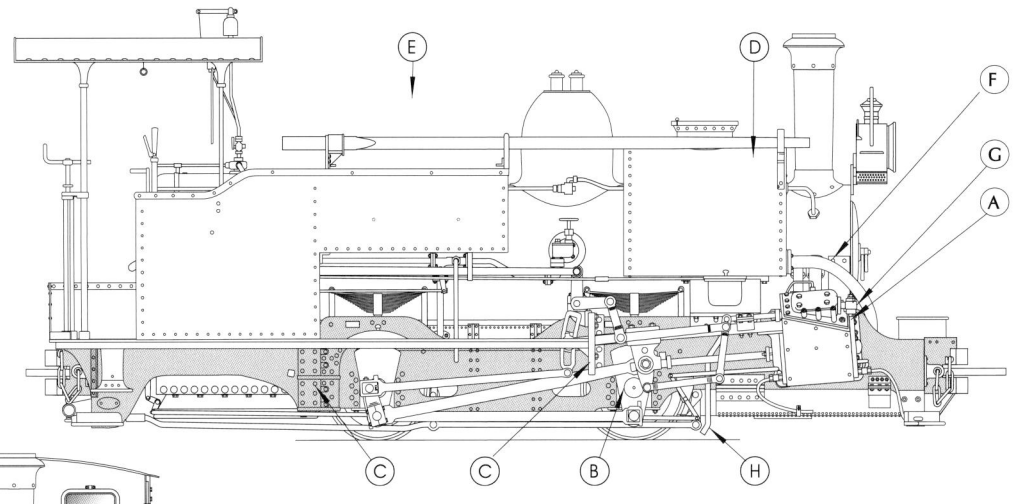

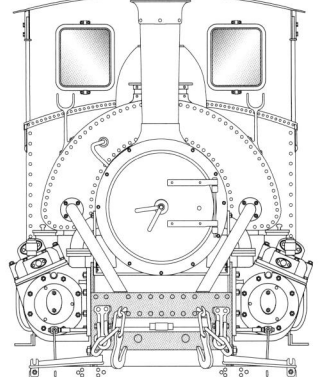

North British 1913 'Revised B' Class as built

- (A) New cylinder / steam chest design.
- (B) Balanced cranks from 1913.
- (C) Bracket where frame widens for firebox and cast motion bracket.
- (D) Longer saddle tank.
- (E) Wakefield mechanical lubricator on left-hand footplate.
- (F) Connecting pipes from saddle-tank join at 'T' before entering well-tank.
- (G) Vacuum relief valve on steam chest.
- (H) Sandpipes for sanding operated from cab.

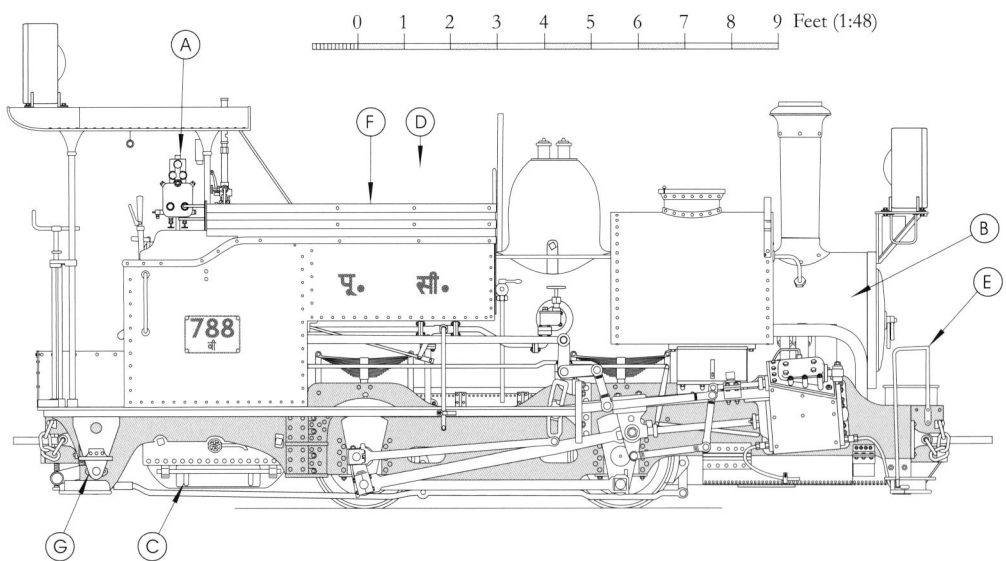

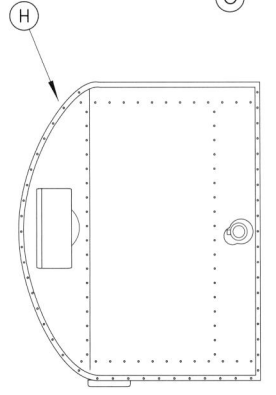

0 1 2 3 4 5 6 7 8 9 Feet (1:48)

788 (Built 1913) as running in 2006

- (A) Cab mounted Detroit sight feed lubricator.
- (B) Extended smokebox introduced from 1920s
- (C) Side door to ashpan introduced from about 1930.
- (D) Turbo-generator mounted on right-hand footplate.
- (E) Sander's handrail added from 1940s.
- (F) Coal rails introduced from 1950s
- (G) New footstep for extended brake shaft from 1960s brake trials.
- (H) Profile of cutaway rear of cab roof introduced about 1968.

©David Churchill 2018

26

Three more 'B' Class to the same design were ordered on 16 July 1913 (order L586), for delivery in February 1914 as Nos 34, 35 and 36. Although the basic dimensions and outline was unchanged, the numerous changes in the 'revised' design included the following:

- Balanced slide valves and new valve chest covers
- Steam chest vacuum relief valve added
- A changed rear frame design with cast brackets joining the front and rear sections of the frames
- Cast (rather than fabricated) motion brackets
- Balanced cranks
- Some changes to the motion, eg different crosshead, changed union link
- Changes to the saddle-tank length and well-tank profile
- Wing-tank extensions under the cylinders not fitted
- Safety valves changed to the 'Drummond' type and repositioned on top of the dome, with the whistle moved to the the steam turret immediately in front of the cab
- Feed pump not fitted, replaced by a second G&C Type 5 injector

The North British general arrangement (GA) drawing showed a switch to a Wakefield Mechanical lubricator. It seems that this did not last for long in practice, although photographs do exist of them on a 'B' Class on the DHR. These locos were built at North British's Atlas Works. They were substantially heavier than the earlier 'B' Class, with weight in working order as delivered being 15.1 tons.

On 9 July 1915, the DHR directors approved the purchase of two new 'B' Class engines *'to take the place of two existing engines no longer serviceable for work on the hills'*. It was left to the discretion of the London agents whether to purchase immediately or *'whether they should wait a little on the chance of prices improving'*. Eventually, an order for two 'B' Class locos was placed with North British, under their contract No L670 with works nos. 21472 and 21473 allocated. In January 1917 they were described as *'nearly completed but the Ministry of Munitions has stopped all work on them'*. Continued wartime restrictions meant delivery would have been long delayed, and the order was eventually cancelled.

Class 'B$_2$' – An 0-4-2 conversion

The first section of the DHR's Kishenganj branch opened on 16 March 1914 and it was fully opened by 1 May 1915. Running across the flat land, the line was very different from the DHR's main line and the Teesta Valley branch, with less severe curves and no steep gradients. Two 4-6-2 tender locomotives were ordered for the line, but existing DHR engines were used to supplement them during construction. To better suit them to the line, two were converted to 0-4-2, it being noted that the unaltered 'B' Class engine was *'not fit for high speeds'*.

No 34, from the 1914-built second order for the revised 'B' Class, at Kurseong station probably in 1946. This has similar modifications to loco No 33, on page 25, but additionally has the extended smokebox, handrails for the sanders and a pipe running along the right hand footplate edge which is believed to have been for a connection into a tender. This pipe runs into the well tank and should not be confused with the injector overflow immediately in front of it.

The turbo-generator for the electric headlight is mounted on the right hand running plate. *(Narrow Gauge Railway Society Ltd. Library / Unattributed)*

During the financial year April 1913 to March 1914, 'A' Class No 9 was rebuilt with *'its frame lengthened and Bissel Truck fitted'*. No further details or photographs have been uncovered of this rebuild which must have been extensive and included *'Fitted with new boiler with new type of dome and safety valves… Both cylinders, crossheads and valve motion renewed and otherwise reconstructed'*.

A similar rebuild was made in the year ending March 1915 to 'B' Class No 22. As an 0-4-2 it was classified as Class 'B2' and it was referred to as such by Mr Addis, DHR General Manager, who one assumes would have known.

It was proposed to use the DHR 'C' Class Pacifics to work the through services over the whole line, and the converted 'B' Class for a local service. It appears from timetables that the local passenger service never materialised.

The 0-4-2 conversions seem to have been quite short-lived. A 1917 list has No 9 as 0-4-0 so it was presumably back to normal by then. No 22 was still listed as Class 'B2' 0-4-2, with four locomotives allocated to the Kishenganj line (two Pacifics, the remaining 'B2' and a standard 'B' Class).

A DHR loco listing dating from between 1919 and 1923 has No 22 back to being an 0-4-0.

No photograph has yet been found of either locomotive as an 0-4-2, and pictures of both engines in the late 1920s give no indication of any remnant of the extended frames.

0 1 2 3 4 5 6 Feet (1:48)

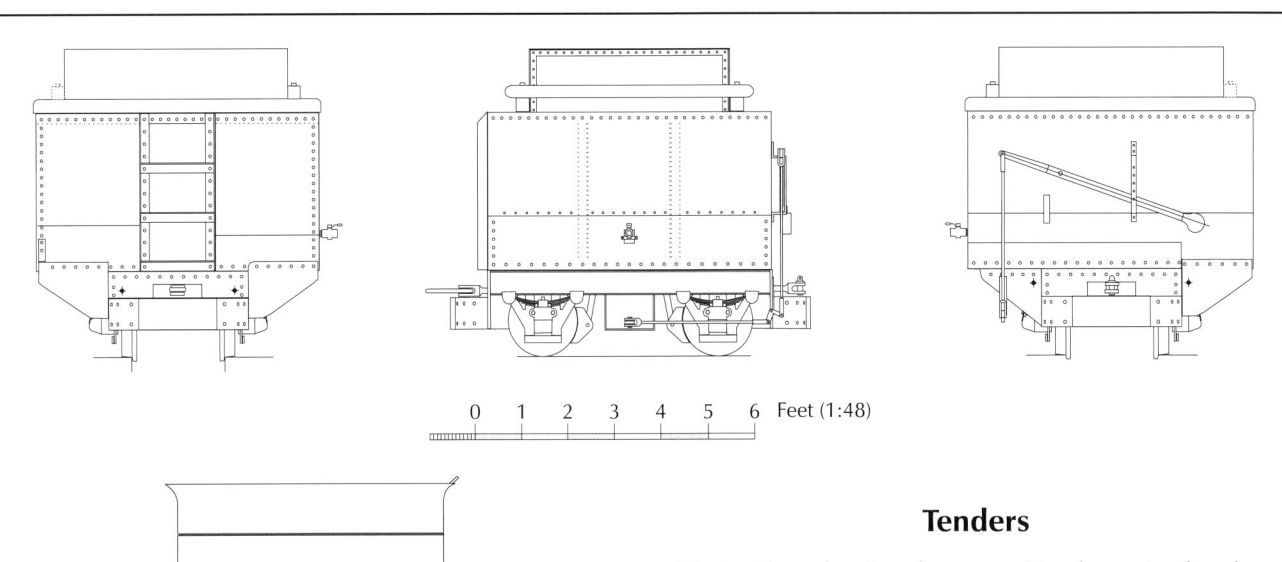

Tenders

The top three drawings show a surviving former tender when running as a water tank in 1979. The build date is not known, but it or a similar vehicle definitely existed in the 1930s and possibly date back to the 1910s. The tank held only 400 gallons and sat within the vehicle body rather than being formed by the body sides.

The drawing [lower left] is a speculative diagram of another type of tender, which ran at the Delhi Centenary Exhibition in 1953 but probably dated from earlier. Either it or a similar vehicle survived until at least 1979. The drawing is scaled from photographs and the length and wheelbase are particularly uncertain.

The red line shows the approximate floor level inside the vehicle, assumed to be the top of the tank, with space above for coal.

Prominent angle along base of side is 3in x 3in approx.

The presence of a short central handrail on one end suggests that the brake lever may also be at that end.

0 2 4 6 Feet (1:76.2)

©David Churchill 2018

Tenders for the 'B' Class

Photographs and a few references, over quite a long period, show that 'B' Class occasionally ran with tenders. Information is sparse and what is known is summarised below.

It is recorded that two 'new tenders' were under construction in December 1910 (the word 'new' may imply that there had been earlier 'old' ones). Several pictures from the period show small tenders on up trains on the main line. They appear to be a plain box on four wheels and it is not known whether they were used for coal, water or both.

Several photos from the 1930s and 1940s show locomotives with additional pipework which appears as if it might be for a flexible water connection to a tender. On the right side of the loco this ran along the footplate edge towards the rear, probably from the well tank, and on the left side from a valved 'T' in the pipe linking saddle to well tank at the smokebox end. No photographs have yet been found of these connected and in use, perhaps understandable if they were primarily for use on the rarely visited Kishenganj line.

The remains of what appears to be one of these tenders survive today. It is now an open box but when photographed as a water tank in 1979 held a 400-gallon tank. This was not a recent addition as it had been pictured in the 1930s, 40s and 50s. Perhaps most significant is a 1940s photo where it was captured in the background in a light-lined dark livery which would match with one of the silver grey locos of the time.

A 1912/13 picture of an up train of early bogie coaches at Sonada with what is assumed to be a tender behind the 'B' Class. Close study of the original suggests a mechanical lubricator may be fitted to the locomotive. See page 89.

No 30, during the 1930s, carried a small tender/water tank. This might be a modification of one of the 1910 tenders above. The engine also has a 'T' and valve visible on the connecting pipe from the saddle tank. *(Bournemouth Railway Club Trust, Kelland collection 50090)*

Water tank TW1 was at Tindharia in 1979. This appears to be the same, or a very similar, vehicle to that shown above. It still survives in poor condition at Tindharia although without the internal tank.

Part of the same or similar vehicle in the background of a 1940s picture [right]. Its use as a tender is suggested by the lined livery which appears to match that used on some 'B' Class at the time. *(Terry Martin collection courtesy Fabien Raymondau)*

A different and rather larger tender was exhibited with 'B' Class No 48 at the 1953 Delhi Railway Centenary Exhibition. This had a very distinctive flared top all round. A couple of this type survived as water tanks through the 1970s, and pictures from above indicate they would be primarily for water.

The streamliner (see page 36) had a tender which appears rather different from any of the previous versions. In contrast to the others, this seems to have been to carry coal. Restored loco 19B, in the UK, now runs with a tender (see page 41).

Built by Baldwin

On 10 July 1916, the directors, while approving a proposal to buy *'three 'B' Class or other types of engines'* instructed the Agents *'to make enquiries whether suitable engines could be purchased from America'*. They approached the Baldwin Locomotive Works in Philadelphia and a detailed Baldwin specification dated 28 October 1916 exists, of which a full transcription is included as Appendix 6.

An order for three 'B' Class locos was placed with Baldwin, based closely on the 'revised' North British design. The specification lists a number of North British drawing numbers and Baldwin's 'Erecting drawing' was very close to the North British drawing for the previous locos. However, there were some changes, partly to align with Baldwin's usual practices and fittings that were readily available. These included 'Consolidated safety valves', two Metropolitan injectors, a Michigan sight feed lubricator, a different type of steam chest relief valve and steam chest drain valves.

They were the first DHR locomotives to be fitted with piston tail rods. Perhaps the initiative here came from the Railway, as the spec notes *'Piston rod extension guide to be as shown on Ry. Co's tracing No2'*.

One of the three Baldwin locomotives showing detail differences. See main text. *(Baldwin Collection, Railroad Museum of Pennsylvania, PHMC)*

Other items mentioned in the specification are:-

> 'On account of the very long downhill runs on which the brakes have to be used continuously, it is necessary to apply the tires [sic] with a very much greater shrinkage than is customary for wheels of this diameter for ordinary service. Give this special attention'.
>
> 'Brake to be operated by vertical shaft at back of cab. Le Chatelier brake valve on line of lowest reading of gauge glass'.

The vertical shaft is the normal handbrake column. The mention of the Le Chatelier brake is a puzzle. It was on the first DHR locos, but appears to have been given up long before this order.

It is interesting that these 'B' Class engines were built at the same time as some of the many 4-6-0 tanks for the British War Department and it seems that they shared some fittings.

The engines were delivered by October 1917 and on 11 October the DHR's directors received reports on them from the General Manager and District Loco Superintendent.

A study of the works photograph [above] shows other variations from previous 'B' Class: for example the injectors are mounted further back towards the cab and tarpaulin weather sheets and supports were fitted at the cab side as well as the rear. Neither of these seems to have lasted in service. In later years the safety valves, injectors and other fittings were replaced with 'standard' DHR parts. However, sight feed lubricators, either Detroit or Wakefield Eureka, subsequently became the standard on all the DHR locos.

The Baldwin built 'B' Class are recorded as weighing 14.65 tons in working order when built; that is about 9 cwt lighter than the previous North British order.

A noticeable feature of the Baldwin 'B' Class was the large circular worksplate on the side of the smokebox which remained on one into the 1970s. A Baldwin plate is in the Ghum Museum.

Built at Tindharia

Three 'B' Class locomotives were built at the DHR's Tindharia works, No 42 in 1919, No 43 in 1923 and No 44 in 1925. It is believed that these would have incorporated parts from the cancelled North British order of November 1916, as a note in the North British order book states *'part material taken by O G & Co, part kept and note of latter given to Dg Off'*. (O G would refer to Ogilvy Gillanders, the UK branch of the DHR's Managing Agents Gillanders Arbuthnot.)

In March 1919 the Government Inspector said of No 42, *'I was much interested in seeing an engine No 42, which had been entirely built in these workshops and at a cheaper cost than it could have been obtained from England. The engine had just completed its trial run satisfactorily and was in the paint shop to be painted; she is of the same type as the 'B' Class engines working on the line but the smokebox has been extended 10 inches and improved safety valves have been supplied.'* As built, they appear to have been a combination of the old and the 'revised' 'B' Class, together with the new features. They were recorded as weighing, when built, 15.75 tons in working order. The 10 inch smokebox extension was longer than that of the later North British locomotives, which was extended by seven inches compared with the earlier engines.

A noticeable feature of the Tindharia builds which remained for many years, was the flush riveted, or perhaps welded, saddle tank without the snap-headed rivets prominent on all other 'B' Class locomotives. In fact, 796 (formerly No 43), now at Tipong, and 797 (formerly No 44), now plinthed at Lucknow, still have the 'flush' tanks. Engine 795 (formerly No 42) appears to have lost its flush saddle tank at some time in the 1980s or 90s.

No 42 TINDHARIA was the first 'B' Class built at Tindharia. The photo date is unknown, but is probably late 1920s or early 30s. These locos were a mix of old and new – No 42 was the first 'B' Class to have the extended smokebox, the safety valves are on the firebox – whether they are the improved type mentioned is not visible. The whistle is on the firebox, but an older one is still present in the former position on the dome. The cast connection between front and rear frames indicates that loco had the 'revised' style of frame. There seems to be a sight feed lubricator in the cab, but there is also a single Ramsbottom type on the cylinder. *(Bournemouth Railway Club Trust, Kelland collection 50093)*

No 43 KURSEONG – the second of the Tindharia built locomotives. Its condition is similar to that of No 42 in the picture above, although this picture probably dates from the 1940s. There are no front handrails for the sandmen, although there are handles on the lamp support.

Two views of 49, the final 'B' Class built by North British, here seen outside Tindharia loco shed, date unknown but sometime in the 1930s. It carries its number at least seven times – as well as the usual cab side, there are plates on either side of the chimney (these were only on the four 1920s North British locos for the DHR), numbers on the lamps and a stamped number on the re-railing bar.

The turbo-generator is in its first position at right rear of the bunker with the exhaust through the cab roof. Tailrods and raised steam chest relief valves are fitted. By the time of this picture, the blower valve had been moved to its usual position on the right side of the dome and side-raking ashpan introduced. *(Bournemouth Railway Club Trust, Kelland collection 50098 and 50099)*

From North British Queen's Park in the 1920s

Further new 'B' Class were ordered from North British in 1925. An order of 19 March (L804) for one locomotive, was amended to three on 7 April, all to be delivered by the end of July at a price of £2,163 each. The delivery date was not quite met – with the locos being weighed on 28 August 1925. The DHR running numbers were Nos 45, 46 and 47.

Five 'B' Class for the Raipur Forest Tramway (RFT) were ordered shortly afterwards on 8 May 1925. (L809).

The design was slightly updated from the previous North British order; the modifications specified included thicker frames and extended smokebox. Immediately visible externally were:

- *The step down in the cab side sheet was further back*
- *Vertical 'L' section front supports for cab roof rather than the round pillars*
- *The cab roof extended further forward at the front*
- *Different blower arrangement (which did not last long on the DHR locos)*
- *The steam chest relief valves are on raised bosses to allow for guides for the extended valve spindles*
- *Various small refinements, eg oil boxes, on some of the pins in the motion*
- *Number plate on the chimney*

These locomotives were slightly heavier than the previous ones, weighing 15.55 tons in working order when built.

Another single 'B' Class was ordered for the DHR on 18 May, 1927 (L839), delivery to be in September at a cost of £2,350. This became DHR No 49. (It was photographed at the North British works carrying plate 48B, but this had already been allocated to the first ex RFT locomotive by the time it arrived in India.)

The North British 1920s-built locos (and those of the RFT) remained easily identifiable for many years by the cab side sheet.

No 45 is ready for despatch from North British's Queens Park works. *(Glasgow Museums & Libraries Collections; The Mitchell Library: Special Collections))*

Works photograph of one of the 'B' Class built by North British for the Raipur Forest Tramway in 1925. Originally intended for wood firing the only difference from the DHR locos was the addition of raised bunker rails around the bunker. *(Glasgow Museums & Libraries Collections; The Mitchell Library: Special Collections)*

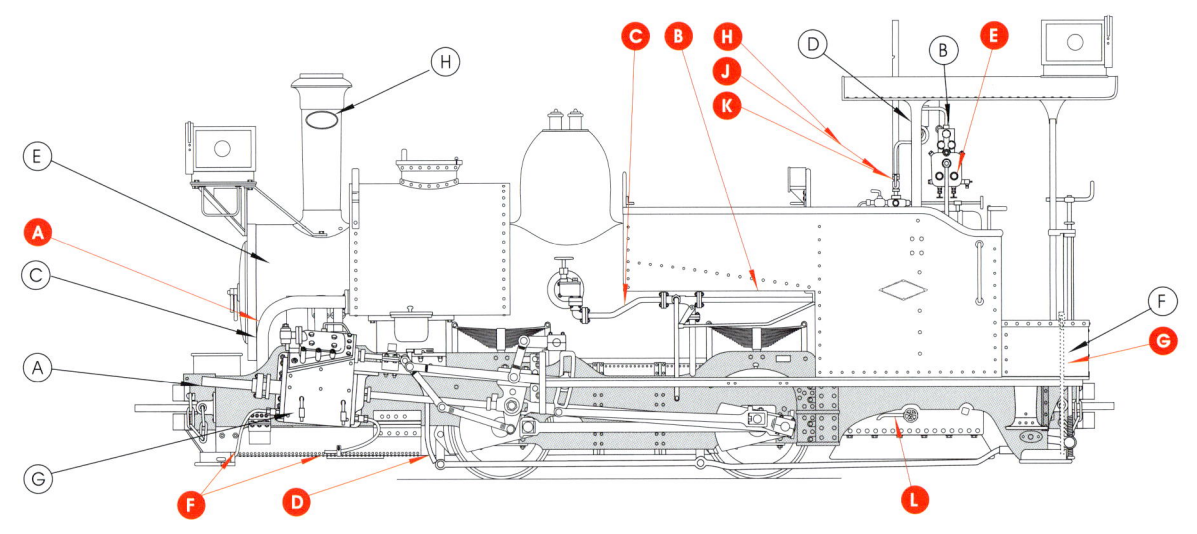

North British 1920s 'B' Class as built

- (A) Tail rods.
- (B) Cab mounted 'Detroit' lubricator.
- (C) Separate pipes from saddle-tank to front of well-tank.
- (D) Angle instead of pillar front support.
- (E) Extended smokebox
- (F) Well tank level indicator / vent (short lived)
- (G) Steam chest auto. drain valves (short lived .)
- (H) Number plate on chimney (lasted until 1950s).

Pipework sizes specified by North British (diameter in inches).

- (A) Tank connecting pipes (2¾in)
- (B) Steam to injector (1in copper)
- (C) Other injector connections ie suction, delivery and overflow (1¼in copper)
- (D) Sand pipes (1¼in steel)
- (E) Oil feeds from lubricator to steam chests (3/8in copper)
- (F) Cylinder drain cock outlets (7/16in copper)
- (G) Pipe for tank gauge (27/32in steel)
- (H) Steam for Detroit lubricator (1in copper)
- (J) Steam for whistle (¾in copper)
- (K) Steam to pressure gauge (½in copper)
- (L) Outlet from blow off cock (1½in copper)

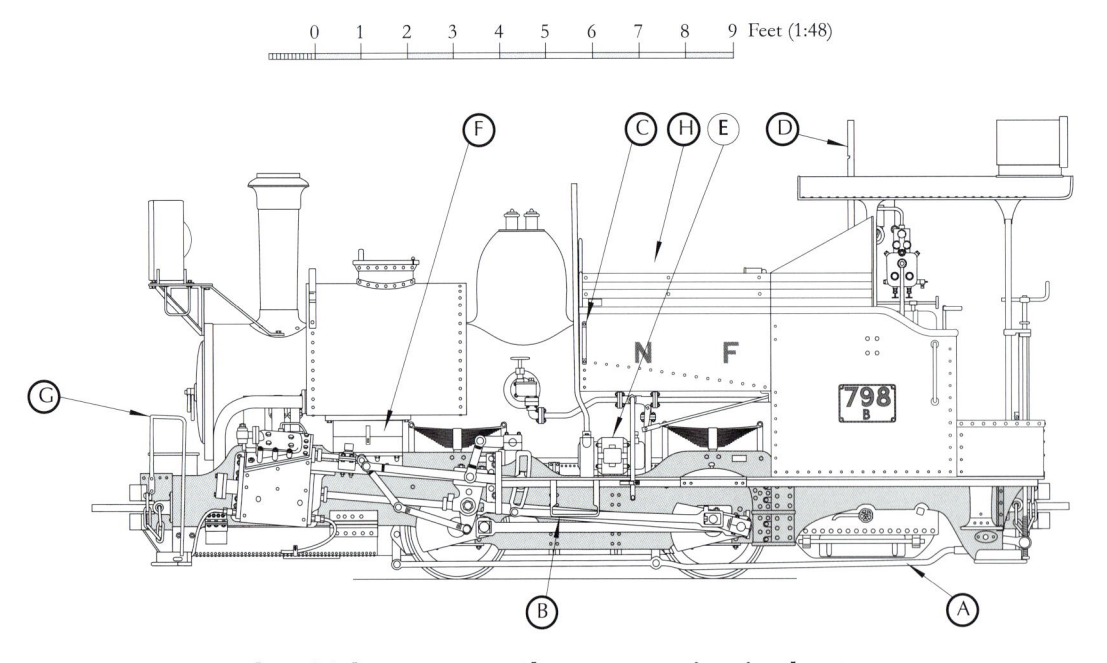

North British 1920s 'B' Class as running in the 1960s

- (A) Side door to ashpan introduced from about 1930.
- (B) Coal breakers footstep and hand hold.
- (C) handrail (unusual).
- (D) Two whistles side by side.
- (E) Turbo-generator mounted on left-hand running plate.
- (F) Sandpipes removed and box changed to toolbox
- (G) Sander's handrail added from 1940s.
- (H) Coal rails introduced from 1950s.

©David Churchill 2018

34

The Raipur Forest Locomotives

The Raipur Forest Tramway (RFT) was a 2 ft gauge line running for almost 68 miles south from Charmuria on the 2ft 6in gauge Raipur-Dhamtari branch of the Bengal-Nagpur Railway. The prime purpose was to carry timber for the Central Provinces Forest Department. It was started in 1924/5 and completed with the opening of the final section to Likma on 27 November 1927. It had the same managing agents, Gillanders Arbuthnot, as the DHR and it is therefore perhaps not surprising that five 'B' Class locomotives were ordered from North British to equip the line.

The order was placed on 8 May 1925 and recorded in the North British order book as being for *'Darjeeling Himalayan Railway'*, soon amended by *'For use on Raipur Tramway – letter 11.5.25.'* They were to be adapted to burn wood with modifications to the firebars and the fitting of a spark arrester.

The North British works numbers were 23300 to 23304, RFT numbers B1 to B5.

One of the RFT 'B' Class, No 2, was transferred to the DHR as early as 1927, where it took the number 48. The RFT was never very successful financially. Its main traffic was wood for railway sleepers, which declined during the 1930s as railways switched to steel sleepers. Much of the traffic was for H Dear & Co, for whom Gillanders were also the managing agents. Some passenger traffic was carried using 'push trolleys', although the official status of this is unclear.

The locomotives were noted as being overhauled during 1937-8 at the Bengal-Nagpur Railway workshops at Nagpur. It appears that this was done regularly when timber traffic ceased for the monsoon season and there is a description of one occasion when they were returned painted apple green and lined out, which was the BNR mainline livery.

After a short life, the RFT finally closed in the year 1941/42. The line's other four 'B' Class became available and were transferred to the DHR in 1943.

RFT NO	B1	B2	B3	B4	B5
DHR number	52	48	51	53	50
Date of transfer	July 1943	1927	June 1943	July 1943	June 1943
Later IR number	805	801	804	806	803

These locos can be easily recognised as they had:-

- *Fuel rails above the bunker sides (not introduced on the DHR locos until the 1950s)*
- *A curved cutaway at lower front of the cab side sheet (not as supplied but soon introduced on the RFT). Photographs indicate that B2 was transferred to the DHR as No 48 without this modification.*
- *Rerailing bars carried further out, necessitating 'splayed out' supports on the saddle tank.*
- *A different arrangement for the blower over the left side of the saddle tank.*

RFT 'B' Class after transfer to the DHR as No 52, probably taken in the 1940s. RFT style bunker rails, curved bunker rear and splayed out re-railing bar support are visible, but the blower pipe appears to run through the saddle tank in DHR style. The turbo-generator being on the right-hand side running plate suggests this picture is earlier than the one below. *(John Thorne)*

The left-hand side of the same locomotive. From the coach livery, the date is probably the mid 1940s. The position of the turbo-generator suggests it is later than the picture above. However, the blower linkage still runs over the left- hand side of the saddle tank – perhaps it is disconnected.

The same loco in 1971, now Indian Railways 805, still retains its typical Raipur bunker curve and splayed-out support *(Michael Bishop)*

For many years the Darjeeling streamliner was thought to be a myth – until this photograph was found! It was certainly a remarkable piece of work *(DHRS Archive)*

[Right] One of the first images to reach the DHRS, and [lower right] a recently-discovered photograph from Diana Hare's family album. Published in *The Darjeeling Mail*, Issue 76.

After their arrival on the DHR, the blower arrangement was soon changed, as were the bunker rails when all the DHR 'B' Class were fitted with bunker extensions during the 1950s. However, the curved cutaway and splayed out rerailing bar support remained to identify ex-RFT locos running on the DHR for many years, until loco parts and identities did eventually begin to get mixed up. Even in 2016, a few of these RFT features remained, although not on ex-RFT locos.

The Streamliner

Rebuilding 'B' Class No 28 as a streamliner with a tender was probably the most bizarre event in the DHR's history. It was unveiled in May 1942 and the rationale behind such an unlikely project remains a mystery. The most likely explanation seems to have been simply to provide a fashionable modern (1930s style) image after the Walford diesel's disappointing performance, maybe primarily for use on the Kishenganj branch and perhaps, as Terry Martin surmises, to act as a morale booster for weary soldiers.

However, whatever the justification, it was clearly a significant engineering exercise, not a quick 'lash-up'. There are still many questions unanswered, but after study of the

The Streamliner 1942

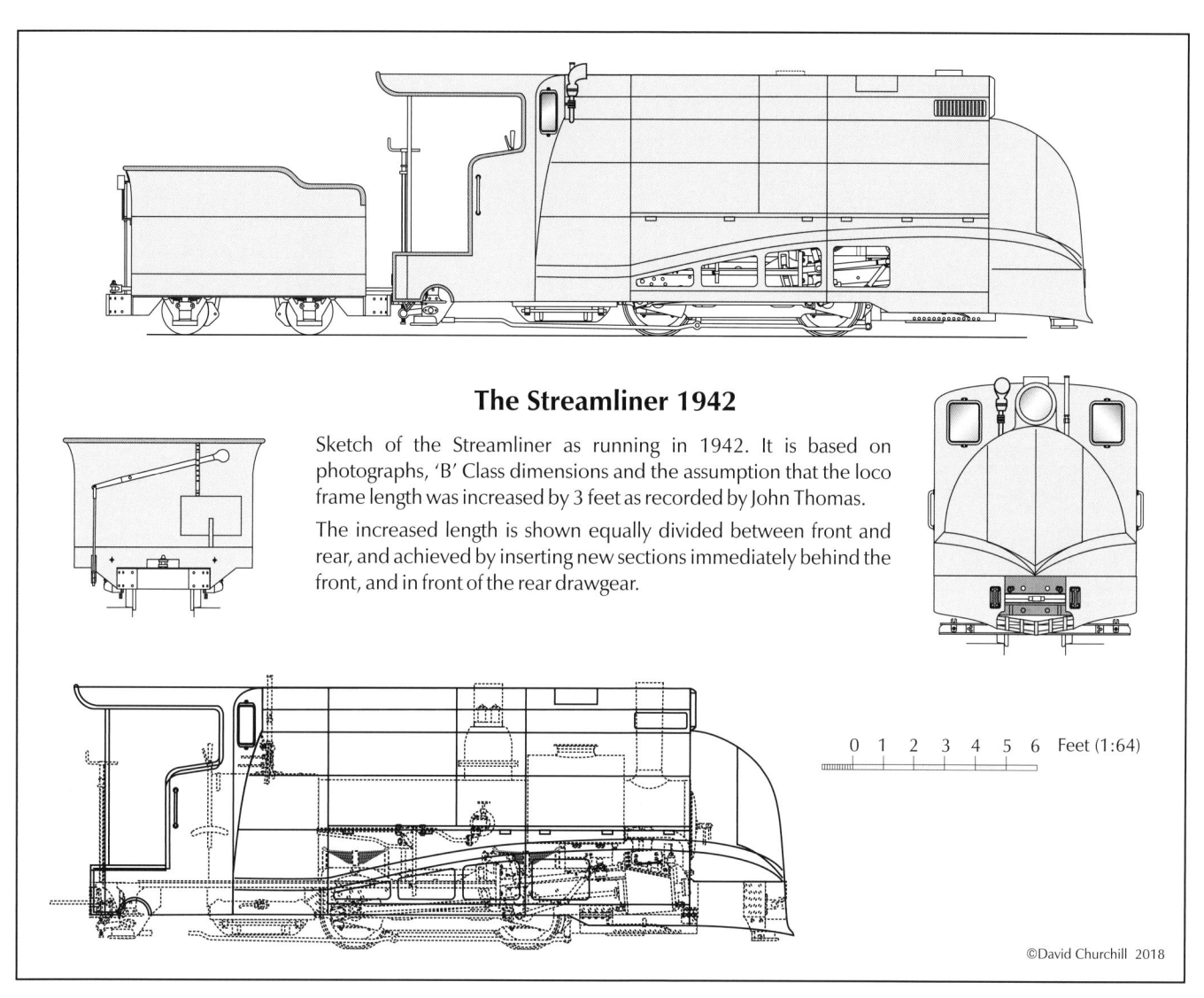

Sketch of the Streamliner as running in 1942. It is based on photographs, 'B' Class dimensions and the assumption that the loco frame length was increased by 3 feet as recorded by John Thomas.

The increased length is shown equally divided between front and rear, and achieved by inserting new sections immediately behind the front, and in front of the rear drawgear.

©David Churchill 2018

known photographs and an attempt to construct a new drawing of the loco, it is now suggested that:

- The locomotive was definitely lengthened (contrary to what Terry Martin believed) at both the front and the rear. Whether this was by the three feet as quoted by John Thomas is not certain, although he is usually a very reliable source.
- The cab was longer and angled at the front. In fact it appears that the whole cab and bunker section was replaced – the pillars supporting the rear of cab roof were quite different to those seen on any other 'B' Class.

The boiler was not the original from No 28 as it has a Belpaire firebox. In addition, the rear view clearly shows a handle on the left-hand side for the linkage to the smokebox blower valve, exactly as used on the 1920s North British locos and those on the RFT. It would therefore have had dome-mounted safety valves, and the side view photograph has a hint of there being an outlet, presumably via a large diameter pipe, just visible in the correct place in the top of the casing.

The device on the right-hand side immediately in front of the cab was simply a steam siren, not common on locomotives but sometimes used on traction engines. This would have given a very effective audible warning when needed.

The rear view shows that there was room for a useful quantity of coal on the footplate, although clearly not enough to replace the bunker. Whether the tender was for water or coal or both remains a puzzle. The locomotive rear view also shows no obvious sign of connection for water supply. The rivet pattern on the tender side may indicate a sloping floor, suggesting it was for coal. It may have been for coal needed on the uphill run but, assuming the loco was not turned, they could manage with just the footplate when running downhill. Running downhill chimney first may not have been very safe and might in fact have been seen as contravening the operating rules.

It is assumed that the intention was to sand by gravity from the original sandboxes operated from the cab. If so, it seems that this was ineffective as there is a report of its first train reaching Kurseong with the front of the streamlined casing wired open and a sandman riding inside. Since there are electric lamps there was presumably a turbo-generator fitted somewhere, although there is no sign of one or its exhaust.

How long the streamliner lasted is unclear. It probably ran for a couple of years, although it was definitely back to 'normal' by May 1945, and was reported as having swapped builder's plates with No 30.

37

The 1940s grey livery

During the Second World War, determined efforts were made to modernise the DHR. This included trials with an unsuccessful diesel locomotive and rebuilt coaches of a more modern appearance, painted in a silver grey colour. When the first diesel trial failed some 'B' Class were also painted in the grey livery: known examples are Nos 28, 30, 41 and 47.

Two pictures of No 47, one of the locomotives painted in the light grey livery during the Second World War. This is one of the final 'B' Class built by North British and still carries its number on plates on the chimney. The turbo-generator is mounted on the right-hand footplate. *(Terry Martin collection and D W K Jones, courtesy Mike Jackson)*

An enlargement from a picture of Darjeeling station, circa 1905. The loco is little changed from its condition as delivered. The pipe across the cab roof, running down into the bunker is the supply to the Wells light. The chimney has the plain top carried by some locos in the 1900s, rather than the usual flared cap. The rear railings on the driver's side are filled in, also seen on other locos at that time. The crosshead-driven feed-pump still is in use *(Hugh Ashley Rayner collection)*

The date of the picture is uncertain but it is probably in the 1930s. A Tindharia rebuild plate (1914) is carried in addition to the large Sharp Stewart works plate. A side-raking ashpan is now fitted and a sight-feed lubricator is in the cab. The boiler appears to have an extended smokebox, but it still has the original whistle and safety valve positions.

Updating the 'B' Class – No 22 through the years

As with all locomotives that work for many years, the 'B' Class have been gradually modified and updated through their lives. The older examples were gradually rebuilt to include many of the new features introduced on later orders, for example:

- *Balanced flycranks*
- *Removal of the wing tanks*
- *Replacement of the feed pump with a second injector*
- *Replacement of various types of safety valves with the Ross Pop type, mounted on the dome*
- *Moving the whistle from the dome to the front of the cab*
- *Extended smokebox*
- *Piston tail rods*
- *Detroit (or similar) hydrostatic lubricators on all locos.*

No 22, in the 1940s. The boiler with extended smokebox, now has Ross Pop safety valves on the dome and its whistle on firebox. The wing tanks have been removed. Handrails have been fitted at front for the sanders, in addition to the handles on the lamp support. *(D.W.K. Jones via Mike Jackson)*

No 22 was converted into an 0-4-2 for a while *(see page 27)*, but no pictures of it as such are known. By 1928 it was back to normal but the revised type of valve chest covers had been fitted, with the turbo-generator on the right-hand running plate plus electric lights to enable running in both directions. The turbo-generator exhaust runs vertically up at front of cab. There remains the original arrangement of the whistle on the dome and the safety valves on firebox.

No 22 in 1972, now numbered 780 and with a very impressive load of coal. The coal rails added during the 1950s are a very noticeable change, the turbo-generator is now on the left-hand running plate with its exhaust in front of the bunker. A different style of front lamp in use. The shorter saddle tank fitted to the early locos is very noticeable in this view. No piston tailrods are fitted. *(Michael Bishop)*

39

Other modifications made to the working locomotives included:

- Side-raking ashpans were fitted from the 1930s
- Switch to electric lights with the addition of turbo-generators and associated wiring
- Replacement of the sandbox fronts with plain toolbox lids
- Addition of handrails for the sanders at the front

It is interesting to illustrate this by following No 22 through the years. It was the sixth 'B' Class, Sharp Stewart works no 3883 of 1892. It was part of the second order no E1010 placed in 1892 and delivered in 1893.

[Right top] An interesting view of 780 in 1982, now named QUEEN OF THE HILLS and carrying a Sharp Stewart works plate. Piston tail-rods are fitted again. The cutaway back of the cab roof is clearly seen – introduced to give more clearance with the new coaches introduced in 1967/8. The bracket on the left hand running plate which carried the speedometer gearbox is present although the rest of the equipment has been removed. *(Chris Pietruski)*

[Right] 780 at Darjeeling station in 2007 named both GREEN HILLS and WANDERER. The turbo-generator exhaust is lying in the bunker. A different type of rear lamp is fitted, with the wiring run neatly in a conduit, along the bunker side. The blue colour is noticeably lighter than in earlier years. *(Dave Priestley)*

[Below] 780 at Siliguri Junction shed in 2013. It was the last of four 'B' Class sent to Golden Rock for reconditioning. Despatched in 2009, it returned in Dec 2010 with a new Veesons boiler, new bunker, saddle tank and various other alterations. *(Peter Jordan)*

The 'B' Class – 1948 to the present day

The 1950s and 60s

The DHR was purchased by the Indian government in 1948, primarily to allow its Kishenganj branch to be incorporated in the rapidly-constructed Assam Rail Link. After completion of the Assam Rail Link and closure of its Teesta Valley branch, following devastating floods in 1950, the next couple of decades were very much a case of business as usual for the DHR.

In 1952, four of the older 'B' Class (Nos 17, 20, 23 and 29) were withdrawn and scrapped as fewer locomotives were needed without the Teesta Valley and Kishenganj branches.

The line became part of the North Eastern Railway in 1952 and subsequently the Northeast Frontier Railway from 1958. Except for very limited use of the two Pacifics around Siliguri, the 'B' Class locomotives continued to work all services on the line.

Apart from livery, the most noticeable changes to the 'B' Class were the addition of coal rails round the bunkers to increase coal capacity and the fitting of brass number plates carrying the new 'All-India' numbers from 1957. There were some efforts at modernisation – attempts to fit vacuum brakes, introduction of speedometers and a trial of NDM1 diesels from Matheran, but none came to anything.

No 19B – the one that left India

19B, which was allocated but never carried No 778, left India in 1962 when it was sold to Mr Elliott Donnelly of Lake Forest, Illinois, USA. It later passed to Hesston Steam Museum near Chicago and, since 2003, has been owned by Adrian Shooter and based on his private Beeches Light Railway, near Oxford, England.

Its fascinating history, which is far too involved to be covered here, is fully described and illustrated in the DHRS publication *The Story of 19B* by David Charlesworth.

Four to Tipong: 781, 784, 789 & 796

The Assam Railway and Trading Co. Ltd was incorporated in 1881 and amongst other activities operated a number of coal mines in Upper Assam. Access required several miles of 2ft gauge track, worked by saddle tank locomotives from Bagnall of Stafford. Eventually further locomotives were needed and four 'B' Class locos, surplus to the DHR's needs, were purchased from Indian Railways in February 1970. They were for use at Tipong Colliery, in north east Assam, which is in a remote and highly scenic area; hilly and wooded with a river running at the bottom of a steep valley. They went via Calorex in Calcutta for overhaul.

No 42 – one of the Tindharia-built locos whilst in NE Railway service, probably in the mid 1950s. It still has its flush-riveted saddle tank and Tindharia builders' plate, but no coal rails have yet been fitted. At that time locos were allocated to a regular crew, and their names were painted on the cab side. *(British in India Museum, courtesy Kathryn Marsden)*

A smartly turned out 'B' Class, No 45 pictured in 1959, has not yet been given its all-India number of 798. It is now lettered NF, rather than NE and has the style of white stripes used at that time. The frames are polished and she retains her chimney-mounted number plate. *(P G Dow courtesy Malcolm Dow)*

It has been suggested that they went to Tipong 'still in the old green livery'. This seems unlikely as the pictures on page 43, soon after arrival at Tipong, show that at least two carried NFR lettering and standard Indian Railways number plates: all retained the standard plates for many years.

By 1973, the majority of mines in India were taken over by the government, and Coal India Ltd was formed in 1975 to run all the nationalised coal mines in India. It directly operated North Eastern Coalfields Limited, which ran Tipong Colliery. Over the years, the four 'B' Class were gradually modified.

The locos became gradually more rundown. The two older engines, 781 and 784, were finally withdrawn sometime around 1992. They were cannibalised for parts, became derelict and in fact 781 almost disappeared.

The 1970s and 80s

The 1970s and particularly the 80s were a time of steadily-decreasing traffic. DHR mail services ceased in about 1984, and by 1990 goods trains had virtually finished. In 1988 and 1989 political problems in the area closed the line for a long period, which must have almost administered the coup-de-grace to the railway. During these decades all services were operated by the 'B' Class, although the number needed to maintain daily services gradually reduced. Few changes were made to the locomotives.

777 to the National Rail Museum, New Delhi

The National Rail Museum (at first called the Rail Transport Museum) in New Delhi officially opened in February 1977. No 777, the second 'B' Class built and the oldest survivor, became one of the exhibits and has been there ever since.

Three put up for sale

Perhaps the low point in the history of the 'B' Class came in 1992 when there was talk of closing the line, and the remains of three locos were put up for sale by tender that had to be submitted by 30 November 1992. Cedric Lodge visited Tindharia and made the following comments on their condition:

- *783 Rolling Chassis, boiler separate in dense undergrowth. Missing connecting rods, motion, springs, front headstock. Boiler appeared complete and probably restorable using UK techniques*
- *793 Rolling Chassis, with boiler on its side on top. Missing cylinders, connecting rods, motion, firebox, foundation ring*
- *797 Rolling Chassis, boiler separate in dense undergrowth. Missing steam chest covers, front cylinder covers, connecting rods, motion*

After protests, the invitation to tender was withdrawn and the locomotives were left at Tindharia. The full story of this dramatic period in the history of the DHR was published in *The Darjeeling Mail* Issues 25 and 64.

Eventually all three locomotives were cosmetically restored as static exhibits.

794 to Matheran

'B' Class 794 was transferred to the Central Railway line at Matheran in 2001 to operate steam specials. Commissioned on 7 May 2002, it ran occasionally on charters. 794 underwent a POH at Parel, Mumbai in 2009 and was converted to oil firing at GOC in 2013. Afterwards, it seems to have hardly, probably never, worked until it was resuscitated for trials in late December 2017. Its performance was reported as "poor" and it was suggested that the operation and maintenance cost would be ten times that of the diesel locos on the line.

798 restored for Howrah Museum

798 was refurbished for the new Howrah Museum which opened in 2006. It is housed under cover in a modern purpose built building together with an ex DHR coach No 75.

New millennium. New locos – steam and diesel and Joy Trains

In 1996 Indian Railways issued a Global Tender for three new oil-fired steam locomotives for the Darjeeling Himalayan Railway. There was no response, but when a similar tender was issued in 2000, there were three offers as described in Appendix 4. That from DLM of Switzerland passed the technical selection but foundered on commercial grounds and Indian Railways decided to proceed with development in-house, as described on page 107.

The new millennium saw diesels appear on the DHR, with two of the NDM6 locomotives built by SAN Engineering of Bangalore being transferred from Matheran. No 604 arrived in December 1999 and No 605 in February 2000. After some initial teething troubles they settled down to working the up and down NJP-Darjeeling trains. Two more NDM6 followed in 2006 and another two in 2016, all from Matheran.

It is probably true to say that without them the full-line through service would not have been sustainable due to the declining capability of the steam locos.

No 25 in the NE Railway red livery, after receiving its coal rails but before being renumbered as 782, probably 1950s. This is another example with the crew's names painted on the cab side. *(Das Studio, Darjeeling)*

By 1962, No 786 has gained the Indian Railways number plates, with a rare light blue background, and brass Northeast Frontier lettering that became the standard in later years.

793, one of the Baldwin locos taking water at Pagla Jhora in 1965 while working an up freight. Water is flowing out of the well tank vent / water level indicator. The speedometer display is in position. Note the brakeman's communication cord running over a pulley at the back of the cab roof. *(the late Peter Bawcutt via Peter Tiller)*

798 working an up freight in 1965. Note the two different tone whistles, the umbrella behind front headlamp, the typical oil can on the running plate and the bags hanging from the front of the bunker. *(The late Peter Bawcutt via Peter Tiller)*

1968 at Sukna with a train which has arrived from NJP being divided into sections for the rest of the journey up the line. The far loco is 793, immediately identifiable as a Baldwin build by the large brass builders plate on the smokebox. The near loco is 781, one of those soon to be sold for use at Tipong. *(Lou Johnson www.worldrailways.co.uk)*

794 is in the sidings outside Kurseong station. It is another Baldwin, but it has lost its worksplate. There are some interesting variations of livery when compared with the previous picture. The polished brass bands around the smokebox front are typical. *(Lou Johnson www.worldrailways.co.uk)*

19B – The one that got away

Derailments don't only occur on the DHR. 19B off the track while on Donnelly's railway at Lake Forest. Note the bell and the air pipe connector at the front. *(DHRS archive)*

The boiler data plate on 19B. The build date of 1889 makes it the oldest working steam locomotive boiler in the world.

19B shown in action on the Beeches Light Railway. It runs with a purpose-built tender carrying, in addition to the coal supply, the equipment needed for air and vacuum braking to allow for its use on other railways in the UK and Europe. 19B also has numerous other additions, including air operated bi-directional sanding, mechanical lubrication and a raised cab roof.

The Tipong Colliery locomotives

Ex-Indian Railways 796 at Tipong in October 1970 in company with DAVID, built by Bagnall as its works No 2134. *(Narrow Gauge Railway Society Ltd. Library/M G Satow)*

781 also in October 1970. It was noted as having 161 pounds on the gauge: its working pressure is only 140 lbs psi! *(Narrow Gauge Railway Society Ltd. Library / M G Satow)*

When photographed on 31st March 1978, externally 789 was still in roughly ex-Indian Railways condition, although rather dirty. She is here accompanied by Bagnall SALLY *(Laurie Marshall)*

47

By 1978 the other three locos had been much altered by overhaul. 781, pictured on 31st March 1978, has had the Ross Pop safety valves replaced by something that would look more at home on chemical plant, necessitating losing the dome cover. The chimney has been replaced by a stovepipe of enormous diameter, there is also a new flat smokebox door and sundry extra external pipework *(Laurie Marshall)*

The other two 'B' Class (789 and 796) remain at Tipong. On 5th March 2008, 789, by now in a light green livery, is seen running towards the adit. Having worked intermittently the colliery is reported (11/2017) to be still producing a little coal but all removed by truck. The railway is currently not operating even for charters. *(Peter Jordan)*

In 2004 789 was being watched carefully by an armed guard while shunting in the colliery yard. The vehicle on the left is the railway's sole passenger-carrying equipment, usually referred to by visitors as the Pullman car!

[Above] What remained of 781 at Tipong in 2004 *(Peter Jordan)*

In 2010 the remaining parts of 784 were combined with something of 781, perhaps only the number plates, to produce a static display loco for the new Heritage Park at Tinsukia. It has the plates for 781 but as it appears to be largely 784, it should perhaps strictly be recorded as being 784 renumbered as 781. It is shown in its novel elevated display position in February 2012 *(Paul Whittle)*

And what was left of 784 with an interesting chimney, also at Tipong in 2004 *(Derek Pratt)*

49

On 22 November 2004, 796, the other B class still at Tipong, posed outside the loco shed. *(Derek Pratt)*

796 with a short train of empties crosses the viaduct on 11th Jan 2010. It is interesting that the vertical screw handbrake has been removed, and other pictures show that all the brake gear has also been removed. Without the Detroit lubricator, this is a steam locomotive reduced to its most basic elements. *(Peter Jordan)*

Back on the DHR itself

A portrait of 795 in 1971. The complete speedometer system – bracket, gearbox, cable and display – is in place, as are the vertical strips to the cab roof. This was one of the Tindharia builds and still retained the flush riveted saddle tank and Tindharia works plate *(Michael Bishop)*

780 is running downhill on a damp day in 1971. The tarpaulin sheet which is generally rolled under the roof is in use to provide very necessary shelter for the crew. Note the loco number is printed on the sheet and there is a 'window' to allow a much needed view for the brakeman. *(Michael Bishop)*

Between Sukna and Siliguri, 782 paused on the flat lower section of the line south of Sukna, while working a goods train towards Siliguri on 5 December 1979. The crew take the opportunity to enjoy a snack. *(Graham Harrison)*

790 in 1976. This loco was in a very dark, almost black, livery for a time during the 1970s. It had been used during the vacuum brake trials during the 1950s and 1960s, and for a time had very large sandboxes either side of the saddle tank. Although many of the modifications had been removed by 1976, it still retained the forward mounted brake shaft. *(Graham Harrison)*

779 running downhill in 1979. The crew are an interesting study – the driver is seated and brakeman ready for action. It then had the 'correct' bunker sides for an early loco, carried its builder's plates and was named MOUNTAINEER.

[Below] 804 rounds Batasia loop in 1979. This was one of the former Raipur Forest Tramway locomotives and still has the characteristic cutaway bunker side and splayed-out rerailing bar support. It also has the triangular footstep left over from the vacuum brake trials of the late 1960s. *(Chris Pietruski)*

In March 1980, 788 is taking water at Tung while working one section of the 3D up train. At that time there was still regular working of trains in several sections.

Early on a dull and misty morning in March 1980, 779 MOUNTAINEER is working hard as it climbs out of Darjeeling with the 7.00am 4D to NJP. The crew are again interesting; the sander is working from the footplate and using the removable platform to hold his sand supply. At that time four bogie coaches was the normal load.

Although the number of road crossings has been greatly reduced since the early days, there are still over 100 on the line, many with very little visibility for traffic. In 1984, during a period when the coaches were in a blue and white livery, 794 bursts across the Hill Cart Road with an up train. *(Chris Pietruski)*

53

An everyday lunchtime sight at Kurseong on 29 February 1984. 798 [left] has arrived with train 2D from Darjeeling. 799 [centre], having arrived from NJP with 3D, is running back round its train and will go to Kurseong shed for servicing. Following this manoeuvre, 798 will move onto the platform road [right] to take over train 3D for its onward journey up to Darjeeling *(Chris Pietruski)*

The DHR has always been susceptible to landslips and on 1 March 1984, 794 picks it way carefully over a recently restored section of line with the school train. The locomotive still has its Raipur features although the front rerailing pole support on the near side saddle tank has broken off. It has a nicely polished chimney cap although the paintwork is well worn and stained *(Chris Pietruski)*

On 29 February 1984, 786 at New Jalpaiguri shed has letter 'S' on the cab side to signify that it was to be restricted to the 'shuttle' service which then ran between NJP to Siliguri Junction. Four such narrow gauge 'shuttles' were timetabled daily in 1981. It also has NJP on the cab side to indicate its home shed *(Laurie Marshall)*

In 1973, 777 is in the sidings outside Kurseong station while still active in DHR service. Speedometer and drive are still in position *(Michael Bishop)*

The rolling chassis of 783 as it was at Tindharia works in 2001. *(Andrew Young)*

On 20 November 1974, 777, as prepared for the new Railway Museum, was moved out of Tindharia works for a sequence in the superb BBC television documentary *'Romance of Indian Railways'* which followed Mike Satow's work in identifying exhibits. His first choice for the museum had been 779 with 777 to remain on the DHR. *(Laurie Marshall)*

Until recently, 777 has been displayed in the open and has needed frequent repaints. It is shown here in 2015 after an excellent restoration by museum staff in a colour believed to approximate to the green used by the DHR. It is now housed under a canopy which will hopefully help reduce the rate of deterioration of the paintwork. *(Subhabrata Chattopadhyay)*

Like many locomotives in India, 783 was restored for display as a static exhibit at Tindharia in 2007. It is now plinthed at the Rail Coach Factory, Kapurthala, Punjab and was photographed there in February 2016 *(Nick Fitzgerald)*

56

Pictured in 2015, 797 is now surrounded by an immaculate garden at the Research Design and Standards Organisation, Manak Nagar, Lucknow. This was one of the Tindharia built locos and still has a 'flush' saddle tank – whether it is original is not known. *(Nick Fitzgerald)*

The third of the locos offered for sale was 793. Restored, it is now an exhibit at the Chennai Rail Museum, which is at the Integral Coach Factory, Perambur. Photographed in 2013. *(Nick Fitzgerald)*

57

The Matheran Hill Railway

[Above] 794 has been transferred to the Matheran Hill Railway near Mumbai. When still coal fired, 794 was photographed shortly after leaving Neral, with a charter on 5 March 2006. Although at 19 Km long it is much shorter than the DHR, this line has similarly sharp curves and severe gradients (60 ft radius, 1 in 20 ruling gradient) *(Peter Jordan)*

[Right and page 59 top] After conversion to oil firing, 794 spent its time in Neral shed where it was not easy to photograph. These two pictures were taken on 5 July 2015. Note the extended rear footplate and the coupling to suit Matheran stock *(Nick Fitzgerald)*

798 was refurbished for the then new Howrah Museum, in Kolkata, which opened in 2006. It is housed under cover in a modern purpose-built exhibition hall together with ex-DHR coach No 75, although that is finished with wood panelling totally unlike anything it would have carried in service. The locomotive is green with red edging, again unrepresentative of anything used when in operation on the DHR. *(Nick Fitzgerald)*

In 1999, what remained of 798 was on the scrap line at Tindharia works, stripped of many of its useful parts. The boiler is NG/13/L of 1926, which was condemned on 18 September 1996 *(Peter Jordan)*

Seen 17 years earlier than Peter Jordan's scrap line picture, the crew appear very concerned about something while at a water stop. 798 was then named GREEN HILLS, the nameplate of which later passed to 780. The loco still has the 'backward' step down on the cab side indicating (correctly) that it was one of the 1920s North British builds: by 1999 this had changed to the earlier type. The loco has piston tailrods and the Raipur-style splayed out re-railing bar support *(Chris Pietruski)*

60

Dampflokomotive G 2/2 für 610mm Spur
Steam Locomotive G 2/2 for 610mm Gauge
Locomotive G 2/2 pour voie 610mm

DLM

Dampflokomotiv- und
Maschinenfabrik AG

This artist's impression illustrates the high tech, Dampflokomotiv-und Maschinenfabrik (DLM) AG, proposal. Looking externally like a 'B' Class, it is fascinating to speculate what might have happened if this had proceeded – perhaps the DHR would by now have become a showcase for 21st century modern technology steam as well as a World Heritage railway! (courtesy: Roger Waller – DLM)

605 and 604 were the first pair of NDM6 transferred to the DHR in 2000. On New Year's Day 2002, 605 is seen passing Kurseong shed while working a down train. Some red paint from its former Matheran livery, can be seen on the engine compartment louvres. 'B' Class 795 is on shed. *(Peter Jordan)*

After the introduction of the diesels, the 'school train' between Kurseong and Darjeeling continued to be steam hauled for some years. On 7 February 2004, 788 TUSKAR in a very dark blue livery is leaving Westpoint with the return service from Darjeeling. Interestingly, 788 was named TUSKAR from 2002 to 2006, as can be seen here. She then carried no name until 2011 when she was renamed, this time with a spelling correction, TUSKER: she carried this name until around 2014/2015 *(Peter Jordan)*

The 'B' Class continued to be used for the Joy Trains, charters and some tourist-orientated services. At Gayabari on 12 February 2006, 804 (left) on a Darjeeling Tours charter passes 788 (right), now TUSKER, which is waiting in the siding on a Jungle Safari train which ran from Siliguri to Agony Point. As there was no passing loop there, another 'B' Class 786 was at the rear of the Jungle Safari to haul it back to Tindharia.

The 2000s saw the continuing rise of the Joy Train, running between Darjeeling and Ghum. In 2015-6 these services provided over 80% of the line's earnings. 791 is shown on 5 April 2013 approaching the intermediate stop which is made at Batasia loop on the outward journey (Peter Jordan)

'B' Class on the DHR 2018

There are 14 'B' Class at present (2018) on the DHR. Most are in working order or can potentially be made so. Following a visit by the Railway Board Chairman's in March 2018, it was announced that all 14 are to 'be made available for use' and fitted with air brakes.

Typical distribution between sheds has been:-

	1999	2004	2012
New Jalpaiguri (closed 2003)	4	-------	-------
Siliguri Junction (opened 2003)	-------	4	3
Tindharia shed	6*	2	0
Tindharia works under overhaul or 'stabled '(i.e. stored).		2	4
Kurseong	2	3	2
Darjeeling	2	2	5

* Both shed and works

The shift of locos to Darjeeling reflects the concentration of steam on Joy Train services in recent years.

Between 2002 and 2010, four 'B' Class were sent in turn from the DHR to Golden Rock works to be reconditioned. They returned to Siliguri with new Veesons boilers and looking somewhat different externally (eg flush sided bunkers and a welded saddle tank on 792) but were not identical. After their return they seem to have had little use on the DHR until they

Close-up of the brass eagle decoration that 779 has carried on her cylinders in recent years. (Nick Fitzgerald)

had received their next scheduled POH at Tindharia a few years later. A 'POH' (periodic overhaul) involves a return to Tindharia which takes place at nominally four-year intervals.

'B' Class no longer on the DHR in 2018

There are now more 'B' Class locomotives displayed, or in use elsewhere, than remain on the DHR. The locos in the UK, at Matheran, Tipong, and in Delhi, Howrah, Chennai and Tinsukia museums have been illustrated previously.

Built in 1892, 779, now named HIMA-LAYAN BIRD, has for many years been the oldest surviving 'B' Class on the DHR. It has been a stalwart of the line frequently used on the Joy Trains and on charters. Here she is shown outside Darjeeling shed on 7 March 2016 *(Nick Fitzgerald)*

On 3 February 2015, 786 is at the south end of the repaired serious breach in the line at Mile 14. This locomotive has, unusually, the Hindi number plate on the 'wrong' [ie left-hand] side *(Peter Jordan)*

782 pauses at Tindharia station while working a charter on 6 March 2018. *(John Buckland)*

804 under rebuild at Kurseong shed on 8 January 2012. It was one of several locos which were overhauled at the other sheds after the landslide damage to Tindharia works and the line below. It is fitted with one of the Veesons boilers which was new in 2006. *(Peter Jordan)*

802 gained the name WHISTLE QUEEN after a competition amongst local children in 2016. She is in an attractive dark blue livery with the decorated cylinder covers which she has had for a number of years. *(Keith Froom)*

791 has been at Siliguri shed for some years awaiting overhaul. Watched over by the local goat, it waits patiently on 9 March 2016. It was still there two years later. It carries two rebuild plates: 'frame rebuilt Tindharia 1963' and 'rebuilt Tindharia 2004'. Although the loco was built in 1913, intermingling of parts over the years means it now has the bunker side profile originally carried by the 1920s North British locos. *(Nick Fitzgerald)*

788 was trapped by motor vehicles at Kurseong shed on 9 March 2016. *(Nick Fitzgerald)*

795, at Mahanadi in March 2016, has one of the Veesons boilers supplied in 2007. Although the Veeson's plate on the smokebox door has gone, it is still readily recognised by the door closed by dogs around the perimeter rather than a central screw. The locomotive had long lost the Tindharia flush-riveted saddle tank and acquired one with the splayed out beam support of an ex-Raipur loco. *(Nick Fitzgerald)*

806 arrived at Golden Rock for reconditioning in September 2006 and was returned in December 2008 It is seen here at Siliguri Junction shed in December 2015. The smokebox door immediately indicates that it has a Veesons boiler. The new bunker side layout differs from that of 792, and visually looks much more like the old arrangement. *(Peter Jordan)*

792 arrived at Golden Rock Works in March 2002 and was returned to the DHR in September 2006. It is shown at Darjeeling in 2016, as reconditioned at Golden Rock with a new boiler, welded saddle tank and flat bunker sides incorporating the former coal rail. It still retained the name HAWKEYE although that was gone by 2018. *(Nick Fitzgerald)*

Another Golden Rock reconditioned locomotive was 805, which arrived there in December 2008, returning to the DHR in October 2009. On 9 March 2017, recently named IRON SHERPA it was under repair at Darjeeling shed. It is interesting to see that it was returned from Golden Rock after rebuild, with the Raipur-style bunker side profile, similar to the one it carried previously. *(Keith Froom)*

780 arrived at GOC in November 2009 and returned to the DHR in December 2010. It was the last of the GOC rebuilds and on 3 March 2015 it was 'flagged off' by the Member Mechanical, Railway Board as the first loco to be out-shopped at Tindharia following repairs to the major landslide at Mile 14. She is seen here at Darjeeling shed on 30 May 2015. *(Nick Fitzgerald)*

In 2002, 787 was the guinea pig for the oil firing trials described in section 10. After these ceased it was left at Siliguri Junction, then moved to Tindharia works and became derelict with many parts removed. It has been reported that it is planned to resuscitate it as coal fired loco No 02, and work was under way at Tindharia in February 2018 *(Peter Jordan)*

After the termination of the oil firing trials the Golden Rock 'new build' No 1001 was converted to coal firing in November 2007. It is seen here on 10 February 2008 shunting at Siliguri shed, still as No 1001 and named HIMRATHI. However it seems to have been little used until it was overhauled at Tindharia in 2013. *(Peter Jordan)*

After overhaul at Tindharia in 2013, 1001 emerged with its number as '01' and named TINDHARIA. It is seen here in the process of being washed out at Siliguri Junction shed on 9 March 2016. *(Nick Fitzgerald)*

More plinthed locomotives

The 'B' Class are popular as display locomotives outside stations or other railway establishments. The first was 799 in Delhi and others no longer needed on the DHR have gradually found new homes after being restored for static display. Most seem to be in fairly secure locations and kept in reasonable condition. Engines 783 and 797 have already been illustrated on pages 56 and 57. Other 'plinthed' locos shown previously are 784 on page 49, 793 on page 57 and 798 on page 59.

With all the 'B' Class accounted for, it was a great surprise when in 2013 another appeared on static display at Kozhikode railway station, near Calicut, carrying No 780B (M) and the names GREEN HILLS and WANDERER. Display locomotives that are fabricated from discarded parts are identified by an 'M' with the number which signifies 'model'. It came from Golden Rock and appears to have been assembled from spare parts from the locomotives refurbished there, and perhaps the new builds. For example the bunker certainly appears to be from the other (real) 780, which had been refurbished at Golden Rock in 2009/2010 and is still working on the DHR.

785 was photographed painted in a very interesting livery outside Dehradun Railway station on a wet day in February 2016. *(Nick Fitzgerald)*

799 is displayed outside the Indian Railways Headquarters Building, Rail Bhavan, in New Delhi. Unlike 785 at Dehradun, it carries 'genuine' DHR livery. Beware, though, photography is not encouraged here! *(Nick Fitzgerald)*

790 is very smart on its plinth outside the Divisional Railway Manager's (DRM) office at Katihar in March 2016. This loco had some unique modifications during the 1950s and 1960s brake trials, but no signs of these remain. However, it does retain the old coal rails, now painted red, yellow and white *(Nick Fitzgerald)*

800 outside Lucknow Charbagh station in July 2015. This loco has the modified rear footstep for the extended brake shaft left over from the 1960s trials fitted on the wrong [ie left] side. Presumably this was a mistake during its restoration *(Nick Fitzgerald)*

801 at the Carriage & Wagon Workshops, New Bongaigaon, Northeast Frontier Railway in February 2016. This has the brake trials modified footstep on its right-hand side, and also a unique arrangement of the handbrake column *(Nick Fitzgerald)*

803 is plinthed outside the DRM's office, Moradabad, Northern Railway. The plaques record that it served the DHR from 1925 to 1997, that average coal consumed was 39 lbs per mile and average water consumption was 45 gallons per mile. *(Nick Fitzgerald)*

There are now two locomotives carrying the number 780. This one appeared in 2013 on static display at Kozhikode (Calicut) railway station, carrying No 780B (M). It was assembled from parts at Golden Rock. The 'real' 780 is still working on the DHR, and has been illustrated on page 67 *(John Browning)*

71

A closer look at details and variations

This section uses photographs and drawings to describe and illustrate some aspects of the construction of the 'B' Class and the many variations and changes there have been. In recent times, particularly, there has been much interchange of parts between locos and identities can be very confused.

Any references to 19B are to the locomotive now on Adrian Shooter's Beeches Light Railway. One of the first batch of 'B' Class locos, it left India in 1962 and retains some interesting features of the early locos. 'Right and left-hand' refer to the side of the locomotive when looking forward from the cab.

Frames

On the first 'B' Class locos the front sections of the frames, from the front buffer back to the throat plate, were of ⁵/₈ in plate, spaced 2ft 11¼ ins apart, with the spacing of the rear frame sections widened by 6 ins to clear the firebox. The connection between the two sections was a riveted angle connection.

From the 1912 'revised' 'B' Class' order, steel castings were used to connect the front and rear frames, and the rear frames, with their front stretcher, were now made from a single piece of steel, rather than being separate pieces. For the 1920s North British locos, the thickness of the front section of the frames was increased to ¾ in.

Wheels

Balanced cranks were supplied on new locomotives from 1912/13. They were subsequently added to all the earlier locomotives.

Tanks – Well tank and 'Wing' tanks

The well tank was supported on brackets on the lower front of the frames and by vertical brackets from the frames between the wheels. It was shaped to clear the axles at the bottom and the reversing shaft at the top. On the early locos the capacity of the tank was 280 gallons (including 20 gallons in the wing tanks). There was an extended front section of the well tank and it appears that this was modified at some time, with the top of the extended section being lowered by a few inches. A short, flanged connection piece was added linking the bottom of the connecting pipe from the saddle tank to the well-tank flange. This can be seen on 19B.

The first twelve 'B' Class followed the 'A' Class in having 'wing tank' extensions of the well tank under the cylinders, although they only added 20 gallons to the capacity. They were omitted from the revised 'B' Class locos, that is No 32 onwards, and were then slowly removed from all the other locos

With the well tank removed, these 'B' Class frames at Tindharia in 2006, show the front frame stretcher, and the rear frames and stretcher in one piece. The rear frames spaced are 6 ins further apart than the front.

Balanced cranks were introduced from 1912/13.

In Tindharia, the frames from 802 in 1999 are in a more complete state, with the well tank, springs and reversing shaft in position (Peter Jordan)

782 is in the shops at Tindharia in 2015 and shows the boiler support in front of the frame stretcher. The two former 'sandboxes' can be seen, which are in fact primarily very heavy cast brackets that support the saddle tank from the frame. The front sandbox has been removed, giving a clear view of the two angled elbows for connecting the well tank to the saddle tank *(see page75)* (Nick Fitzgerald)

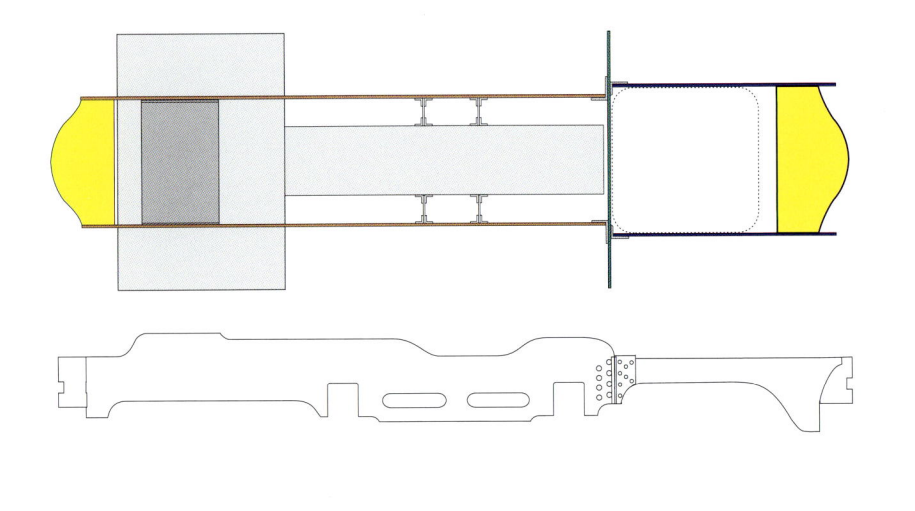

Frame layout

Frame layout and profile of the frame for the Sharp Stewart 'B' Class design.

On these locos the connection between front and rear sections of the frame was by rivetted angle. On the plan view the drawbar/coupling housing assemblies are shown in yellow and individual frame sections in brown, green and blue. For tank details see below.

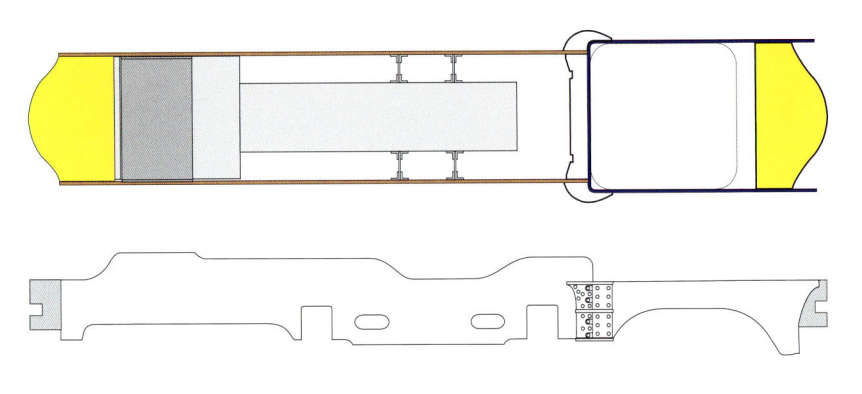

For the North British revised locomotives of 1913 the frame design was changed.

The connection between the front and rear frame sections was now made with cast brackets and the rear frames were made as a single piece. The front drawbar assembly was longer, with the shallow front section of the well tank underneath. The same design was used in the later locomotives.

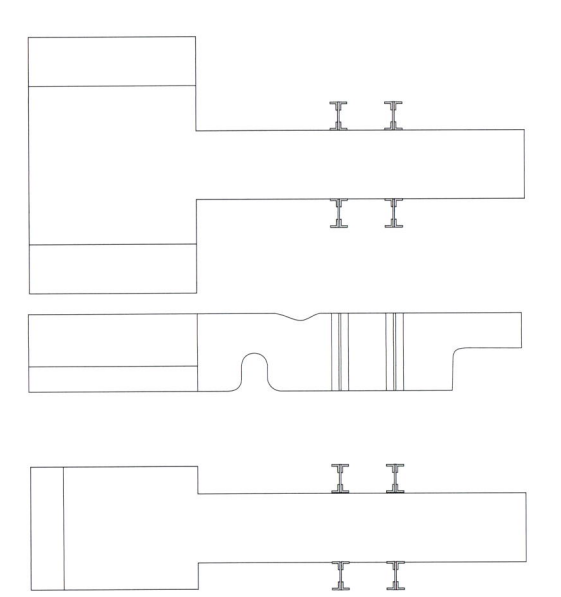

Well tanks

The well tank plan and profile for the first 16 'B' Class (including 2 for Kalka-Simla).

These had the well tank carried forward at full height and extended over the rear axle. They also had the wing tank extensions below the cylinders

The wing tanks were gradually removed from the early locos and, possibly at the same time a section at the front of the tank was lowered by a few inches. The extension over the rear axle was retained.

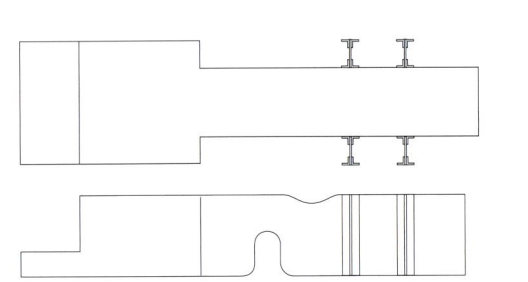

The 'revised 'B' Class of 1913 and later locomotives had a different, and lower capacity well tank. The extension over the rear axle was eliminated and a shallow section introduced at the front which fitted underneath the altered front drawbar assembly.

©David Churchill 2018

in succeeding years, although some remained at least until the 1940s. The surviving remains of 'A' Class No 11 still have something of the wing tanks left (see page 14).

The profile of the well tank was also changed from the 1912/13 revision. The extended front section was removed, also the section from the rear axle to the front of the firebox. The reduced capacity of 230 gallons was accounted for by enlarging the saddle tank to maintain a capacity of 380 gallons. Diagrams of the various tank shapes are included in the drawing opposite.

Saddle tank

The saddle tank is supported from the frames by two very solid brackets that also formed the housing for the original sandboxes. The first 'B' Class locos had a saddle tank 37in long, holding 120 gallons. The 1912/13 revision enlarged the saddle tank to a length of 41¾ in and a capacity of 150 gallons to compensate for their shorter well tank and maintain the total water capacity. The difference between the long and short tanks is quite noticeable in side-view photographs.

The saddle tanks had very noticeable snap head rivets, except for the three locos built at Tindharia which had either flush riveted or welded tanks with no visible rivet heads.

Some of the rebuilds carried out in the 2000s have larger welded saddle tanks. There has generally been a tube running through the tank from back to front on the right hand side to carry the pipe connection to the blower (see page 77).

Pipes connecting the saddle tank to the well tank

The arrangement of the pipes connecting the front of the saddle tank to the well tank varied between the different orders. These variations remain to the present, although locomotive identities are completely mixed up!

Boiler and associated equipment

The 'B' Class boiler had an exceptionally long barrel and a very wide firebox for a four-wheeled, two-foot gauge locomotive. The design has not changed greatly, although there have been changes in details and in materials. Some of the boilers have had a remarkably long life. In fact, that on 19B, dating from 1889, is believed to be the oldest working steam locomotive boiler in the world. A diagram of the boiler is included on page 77. Replacement boilers have been supplied at intervals: details of their suppliers are not yet known. Some had Belpaire fireboxes. I did record an example in Tindharia shed in 1979 with a data plate indicating it came from Semmering Graz of Wien in 1962. (This is not surprising as Austria supplied a number of BG and MG locos to Indian Railways in the 1950s).

The earlier fabricated arrangement to connect front and rear sections of the frames is shown on 19B in 2011. She is now fitted with air-operated sanding, and the sandpipe for reverse running can be seen in lower centre of the picture.

The later design with a cast bracket to connect the front and rear frames on 804 at Siliguri shed in 2006.

A total of eight new boilers have been supplied since 2002, a number from Veesons Energy Systems of Tiruchchirappalli – the oldest on the current DHR loco stock list is recorded as dating from 1928.

Dome and Safety Valves

The early locos had Ramsbottom safety valves mounted on the firebox, with 'chimneys' to vent the exhaust above the cab roof. For the 1912/13 order, Drummond-type valves, mounted on the dome, were used. The Baldwin locos had Coale valves, also dome-mounted. The Tindharia-built locos reverted to the earlier Ramsbottom type, presumably using spare parts available from their stock.

The earlier pattern of well tank lies upside down with the rear of the tank to the left, shaped to clear the front axle. The section of the tank above the rear axle is visible *(Peter Jordan)*

The later pattern of well tank lies on its side with the rear of the tank to the left. The tank finishes short of the rear axle. At the bottom of the well tank are two bolt-on cover plates to allow access to the inside of the tank. There is also a bolt-on rectangular connection between the front and rear of the tank passing underneath the front axle (not in place in these pictures) *(Alan Walker)*

There is a small diameter pipe attached to the front of the bunker which is taken from the lower right-hand side of the well tank with a valve which is left open during tank filling. This bleeds any air trapped in the rear section of the well tank as it is filled and also provides a visual indication when the well tank is full. In this February 2006, the well tank is already full and water is pouring from the pipe as 804 is being intently monitored at Westpoint.

[Above] The tube which carries the blower steam pipe is visible [left] on the saddle tank from 782 in 2016 *(Nick Fitzgerald)*.

[Top right] The detail of the tank filler on 19B, seen here at Leighton Buzzard. This is typical of that on most locomotives.

[Lower right] In contrast, 782 carries a welded fabrication rather than the normal riveted version *(Nick Fitzgerald)*

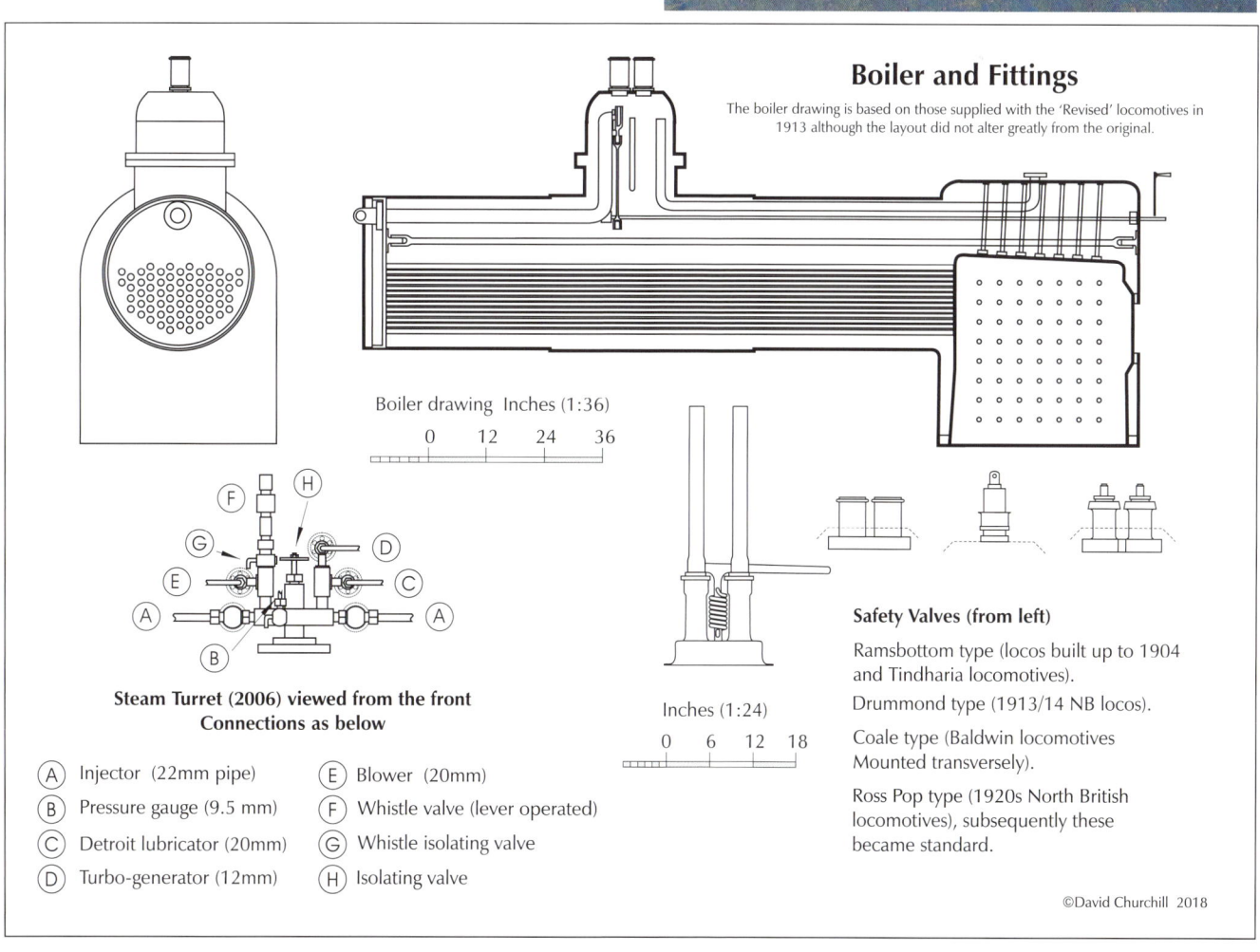

Boiler and Fittings

The boiler drawing is based on those supplied with the 'Revised' locomotives in 1913 although the layout did not alter greatly from the original.

Boiler drawing Inches (1:36)

**Steam Turret (2006) viewed from the front
Connections as below**

- (A) Injector (22mm pipe)
- (B) Pressure gauge (9.5 mm)
- (C) Detroit lubricator (20mm)
- (D) Turbo-generator (12mm)
- (E) Blower (20mm)
- (F) Whistle valve (lever operated)
- (G) Whistle isolating valve
- (H) Isolating valve

Inches (1:24)

Safety Valves (from left)

Ramsbottom type (locos built up to 1904 and Tindharia locomotives).

Drummond type (1913/14 NB locos).

Coale type (Baldwin locomotives Mounted transversely).

Ross Pop type (1920s North British locomotives), subsequently these became standard.

©David Churchill 2018

77

Locos built up to 1904 had separate vertical connections from each side of the saddle tank into the extended front of the well tank.

[Right] The 1912/3 revised 'B' Class had a single horizontal connection into the centre front of the shorter well, linked via a 'T' to each side of the saddle.

The final arrangement (1920s North British) had two horizontal connections into the front of the shorter well connected via a right-angle elbow to each side of the saddle.

The long barrel of the 'B' Class boiler is illustrated by boiler 7/L/NG of 1939 which was plinthed inside Tindharia works in 2010. This example is one of the replacement boilers which were fitted with a Belpaire firebox *(Peter Jordan)*.

One of the new Veesons boilers at Tindharia in March 2008 shortly after delivery. These were installed on 795 and 804. They are readily identified by the smokebox door fixing dogs and handle *(The late Terry Martin, courtesy Fabien Ramondaud)*.

The 1920s North British-built locos were fitted with dome-mounted Ross pop valves, which were gradually fitted to all the locos over the following years.

The safety valves have sometimes been mounted transversely across the dome.

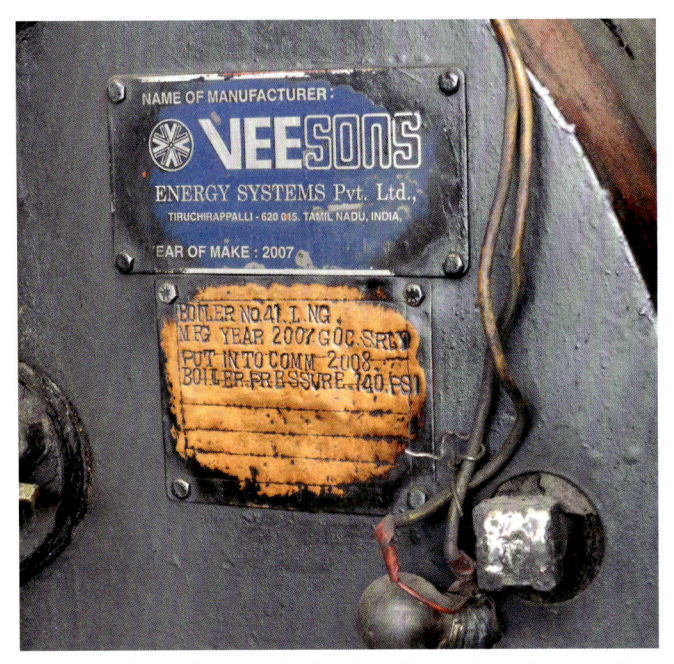

[Above] The makers and boiler identification/test plates on the backhead of 795 in 2015. Note the crude wiring and the simple light bulb *(Nick Fitzgerald)*

[Below] The dome with Ross pop safety valves on 805 in March 2006 and [above] the dome cover.

The injectors

The first seven 'B' Class were supplied with an injector fitted on the left-hand side and a crosshead-driven feed pump on the right side. This combination seems to have been the normal practice at the time. The injectors were at first specified as simply No 5, possibly of Sharp Stewart's own manufacture. After 1899 they were specified as being 'Gresham & Craven automatic restarting No 5'.

The injectors take their water supply from near the bottom of the well tank through perforated pipes which act as filters. Taken during the DHRS Engineering Project, this shows the pipes on 806, viewed through the removed access cover in the bottom of the tank. These injectors had failed due to too great a restriction to the flow on their inlet *(Mike Weedon)*.

A view of the injector typical of those used on the 'B' Class. The steam connection is to the right, the water connection from the well tank is at the bottom via the water control valve. The feed to the boiler is on the left. The blue pipe towards the camera is the overflow, which is taken below the running plate. This pipe originally passed through a hole in the running plate, but was later clipped to its outside.

The injector feed enters the boiler via a clack valve, here shown on 805 in 2006. They are often polished brass.

Ashpan

Clearing the ashpan is a regular task whenever the train pauses to take water or at a station. Until the late 1920s this had to be done, rather inconveniently, through the dampers at the front and rear. About the time the last North British locos were delivered, doors to allow the ashpan to be emptied from the side were introduced. They were soon fitted to all the 'B' Class and must have made raking out at intermediate stops a much easier process.

[Right] Detail of ashpan side door on No 805, February 2006

Clearing the ashpan on a down train, also in 2006. The locomotive 804 is one that was originally at Raipur and still retained the curved cutaway side that was introduced there.

Blower Operation of the blower varied considerably over the years. The table below summarises the changes.

Approx date	Blower valve position	Valve operation	Steam pipe from valve to blower	
First series	Right-hand side of dome	Via rod through bunker	Over saddle tank into chimney base	Altered to be as below – date uncertain, but early
First series later	Right-hand side of dome	Via rod through bunker	To left hand side of loco behind saddle tank, then via tube in tank formerly occupied by blastpipe flap adjuster.	Altered to be as below – date unknown
Locos built from 1899, 1912 revised design & Baldwin	Right-hand side of dome	Via rod through bunker	Through saddle tank into right-hand side of smokebox	
1925 series	Left-hand side of smokebox	Rod on left hand side through bunker & over saddle tank	In smokebox	Short lived except on Raipur locomotives*
Late 1920s	Right-hand side of dome	Via rod through bunker	Through saddle tank into right-hand side of smokebox	1925 series locos at DHR modified
1960s	In cab supplied from steam turret	Direct by hand wheel in cab	Through bunker and saddle tank on right-hand side	Eventually fitted to all locos

* Still present on Raipur locos when transferred to the DHR in the 1940s.

This picture of No 29 derailed, is extremely interesting. It gives an underside view of an early locomotive that still retains most of its early features. The ashpan, with dampers front and rear but no side doors, is quite clear, as are the wing tanks and early pattern cylinders. The earlier fabricated, rather than cast, connection between front and rear frame sections is immediately apparent. Just visible is the rectangular pipe which, bolted under the front axle, provides an additional connection from front to rear of the well tank. This loco also has a mechanical lubricator that is shown. *(DHRS Archive)*

Adjustable blastpipe

The first seven 'B' Class (that is, deliveries before 1900) had a flap top on the blastpipe, operated via a linkage from the left side of the cab. The operating rod passed through a tube on the left side of the tank. Locos built after 1900 had a normal fixed blastpipe nozzle and the adjustment was soon removed from the early locos. However, the through tank tube then seemed to be used to take the steam feed pipe from the blower valve through to the smokebox. This blower arrangement eventually changed but the tube, or signs that it had been there, sometimes remained. Very surprisingly, it is still present on 19B's tank and is now used for the electric supply to the headlamp.

Extended smokebox

This was introduced on the Tindharia-built locos and was fitted from new to the subsequent North British-built locos. It was then gradually fitted to all the earlier locos.

Superheating

There have been at least two trials with superheating. A locomotive was recorded as being fitted with a superheater in 1917. In addition, No 44 is reported (in 1944) to have been tried with a superheater at some stage – clearly not in 1917 as it was not built until 1925.

Both trials were probably short-lived, perhaps due to lubrication difficulties.

Steam turret

The steam turret on the firebox immediately in front of the cab provides a valved steam supply for the injectors, blower, lubricator, turbo-generator and whistle.

Cylinders

The 'B' Class locos supplied up to 1904 had a horizontal top to the valve chest cover similar to that of the previous 'A' Class. This can be seen in pictures on page 19. The remains of 'A' Class No 11 now at Tindharia are the only surviving physical example of the earlier design (see pages 14-15). For the revised 'B' Class from 1912, balanced slide valves were introduced and the top of the valve chest cover changed to slope forward and sideways, as seen on all the 'B' Class locos today. A steam chest vacuum relief valve was introduced at the same time. The later Baldwin, Tindharia and North British builds all had the revised pattern cylinders, and by the 1940s the earlier locos had been gradually changed to the newer design.

Piston tail rods

These were first introduced on the Baldwin locos of 1917, and then fitted to Tindharia-built and 1920s North British-built locos. They were subsequently put on to the earlier locos. However, they were later removed, then re-introduced on some locos. At present about half of the locos on the DHR have them.

The steam turret on 806, which was the 'B' Class overhauled during the DHRS Engineering Project [see 97]. Mike Weedon said: *'It took half a day of digging out coal dust to reach it! The blower valve had come apart and had to be rebuilt before we could test run the loco.'* From the bottom, the connections are to the injectors, blower and lubricator, turbo-generator and whistle working upwards *(Mike Weedon)*

The left-hand cylinder of 791 in 2013 shows the sloping valve chest cover introduced from 1912. The steam chest pressure relief valve is fitted to the front of the cover, the brass front right-hand cylinder drain cock is at the bottom, and the feed from the cab-mounted hydrostatic lubricator at top centre. The loco is fitted with piston tail rods, their housing extending forward from the front of the cylinder *(David Barrie)*

Operation of the drain cocks

The cylinder drain cocks are operated from a lever on the right hand (driver's) side, via a linkage that was originally low down near to the brake pull rod. The layout is clear on the photograph on page 85, where they have helpfully painted it white. The operating lever position high in the cab cannot have been too convenient. The linkage was later changed to pass above the reversing shaft from an operating handle low down on the right-hand side.

Motion brackets

Until the 1912/13 order, the motion brackets were fabricated from plate and angle section. After this, they were steel castings. Earlier locos were modified gradually. No 19B, at least, still has the fabricated bracket.

Crosshead

The crosshead was altered with the design revisions of the 1912 order. The revised design had the lower part cranked outwards to allow a forked connection for the union link. The earlier locos gradually acquired the later design but the older type still survives on 19B, although it is not the original, which would have had a connection for the drive to the feed pump.

Valve gear

Walschaerts valve gear was introduced with the 'No1' Class and its use continued on to the 'B' Class. The basic geometry on the 'B' Class has barely changed since 1889, although there have been numerous detail changes. Study of the works photographs will highlight some of the differences. The North British 1920s locos were supplied with an oil box that allows a greater supply of oil to the oscillating pins than the usual oil hole. The oil box was mounted on the end of the pin and fed through the pin to the joint. It is shown on the works photograph on page19. Examples can be seen at both ends of the lifting link and at the joint between the

View of the cylinder with steam chest cover removed to show valve. This is on the Golden Rock new-build, being worked on at Siliguri Junction shed in 2009 when it was still numbered 1001. The curved cover over the steam pipes exists only on this engine *(Ian Fraser)*

radius rod and combination lever. They seemed to have disappeared over the years, probably because it was easier to replace them with normal pins.

Lubrication

The early locomotives had very prominent Ramsbottom displacement lubricators for the cylinders and valve chests.

The North British GA drawing for the 1912/13 order shows these being replaced by Wakefield Patent No1 pattern mechanical lubricators (four-feed ½ gallon capacity) mounted on the left-hand running board and driven from the motion. It seems that these were short-lived. There are also pictures showing mechanical lubricators fitted to the earlier locos Nos 29 and 30. This may have been a trial as the linkage used does appear to permit to allow the possibility of a considerable amount of adjustment.

On 786 in 2006 they had obligingly painted most of the drain cock linkage white (together with that to the injector water valve and the reversing lever). The linkage operating the drain cocks can be seen winding its way through under the bunker and then over the reversing shaft.

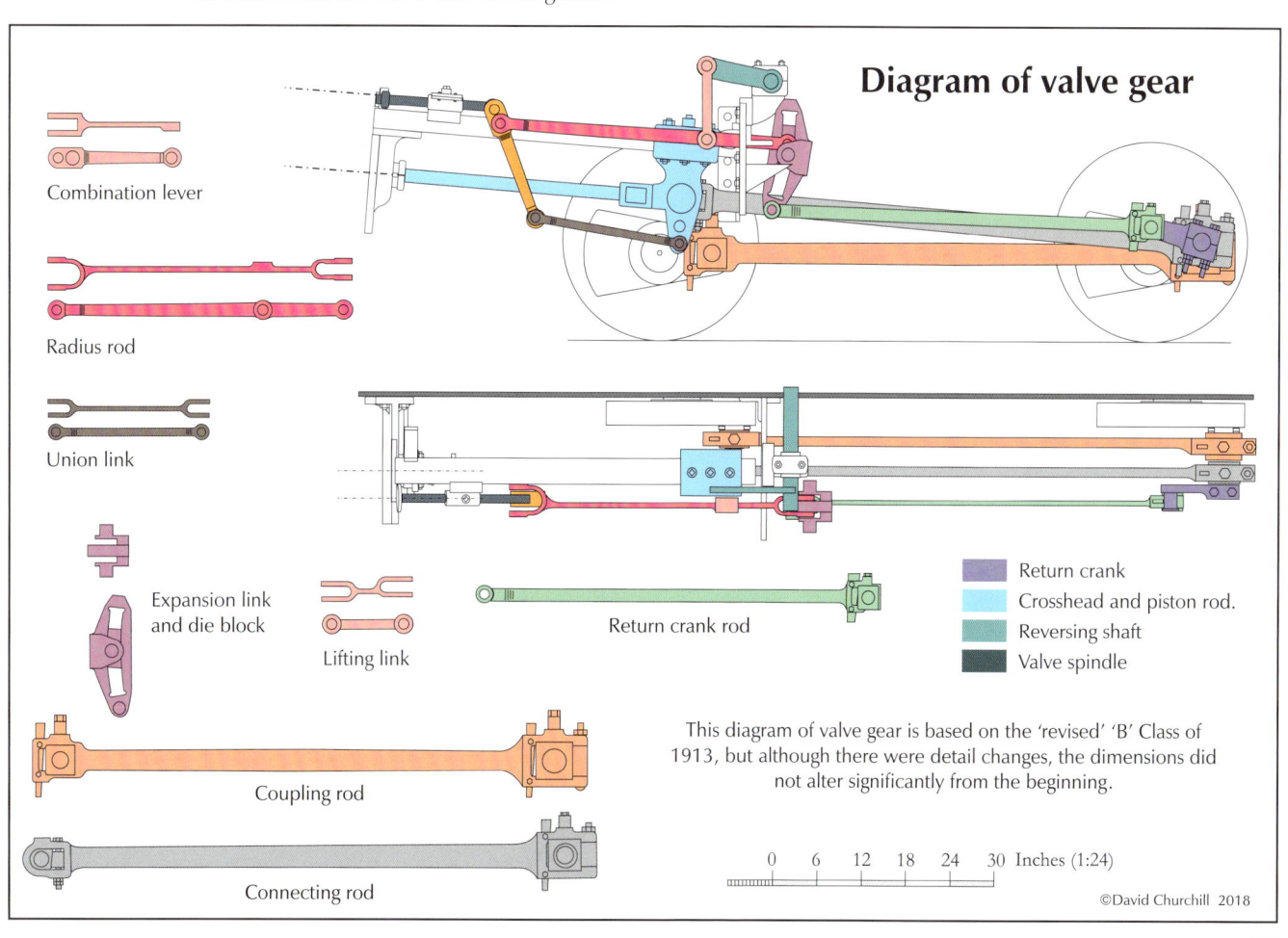

Diagram of valve gear

Combination lever

Radius rod

Union link

Expansion link and die block

Lifting link

Coupling rod

Connecting rod

Return crank rod

Return crank
Crosshead and piston rod.
Reversing shaft
Valve spindle

This diagram of valve gear is based on the 'revised' 'B' Class of 1913, but although there were detail changes, the dimensions did not alter significantly from the beginning.

0 6 12 18 24 30 Inches (1:24)

©David Churchill 2018

85

The Baldwin locos were supplied with a 'Detroit' hydrostatic lubricator fitted on a pillar on the left side of the cab and this type has been used since, early locos being fairly quickly updated. The 1920s North British locomotives were supplied with Detroit four-feed lubricators. Wakefield Eureka lubricators seem to be most frequently used, usually of the two-feed type.

Cab, bunker and roof

The overhang of the cab roof at the front was increased by 2¾ in from the locos supplied from 1925.

During the late 1960s, the profile of the roof was rounded off at the back to provide more clearance when used with the new coaches introduced in 1967/8. This was done very quickly for all the locos in use. No19B escaped this, having already departed to the USA.

Cab front & cab roof supports

On the earlier locos both the front and rear of the cab roof were supported on circular pillars and the cab front sheet had a section in the centre shaped to clear the safety valve base. From the 'revised' locos of 1912/13, the cab front was flat and from the 1920s North British locos, angle sections were used for the front supports.

Coal bunker and coal rails

The rebuilt 'A' Class locos carried coal in a bunker in front of the cab, and this continued with the 'B' Class. The left side of the bunker had a sloping base to help pass coal back to the fireman. This is noticeable because of the sloping line of rivets on the bunker side. The bunker was rated to take 15 cwt of coal, although many photographs show a very high and neatly packed coal stack, suggesting that substantially more was often carried.

The profile of the cab side changed from the 1925 North British locos with the 'step-down' for the cab side moved further back. This feature remained, allowing easy identification of the later builds, until loco identities gradually became confused in later years.

The Raipur locos were fitted with rails above the bunker sides, originally for wood fuel. These were retained for a while after their transfer to the DHR on at least some locos, perhaps until the extensions were fitted generally in the 1950s. Another distinctive feature of the ex-Raipur locos was the curved cutaway section at the lower front of the cab side, possibly to ease access to the washout plugs or perhaps to ease 'flow' of the fuel when they were wood fired.

From the mid-1950s, extensions were added to the top of the bunker sides on all the remaining 'B' Class locos to increase capacity to a nominal value of 21 cwt. A drawing dated 1954 shows an extension around the bunker area only, not extending back to the cab front. Apart from the Raipur locos, no picture has been traced of a locomotive fitted like this, so it may never have been done in practice. The exact shape and detail of the bunker extensions has varied somewhat – examples can be seen on pages 93 and 94.

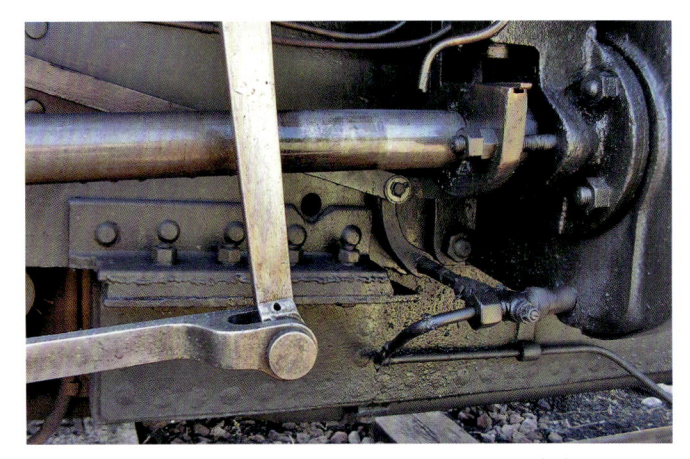

Close-up of a rear drain cock on 19B. The operating linkage seen in previous picture enters from top left and operates the valve attached to the cylinder at lower right. The link to operate the front right-hand cock runs through a recess at the back of the cylinder casting to operate a similar valve at the front. A link passes through the frames to operate the cocks on the left-hand cylinder.

19B at the Beeches Light Railway in September 2012 shows the earlier style of fabricated motion bracket. In the background is the turbo-generator now fitted to 19B, which is from Eastleigh and is of the type used on Bullied Pacifics.

In Tindharia works – February 2006. This loco has the cast motion bracket, with a bearing for the reversing shaft bolted to it. The reversing shaft shaped to clear the bottom of the boiler is visible as is the recess in the top of the well tank to clear it. On the extreme left is the end of one of the angles that support the sandbox and saddle tank. Note also the footstep for the coal breaker attached to the edge of the running plate.

Earlier type of cross head on 19B

The later type of crosshead on 792 in May 2015 *(Nick Fitzgerald)*

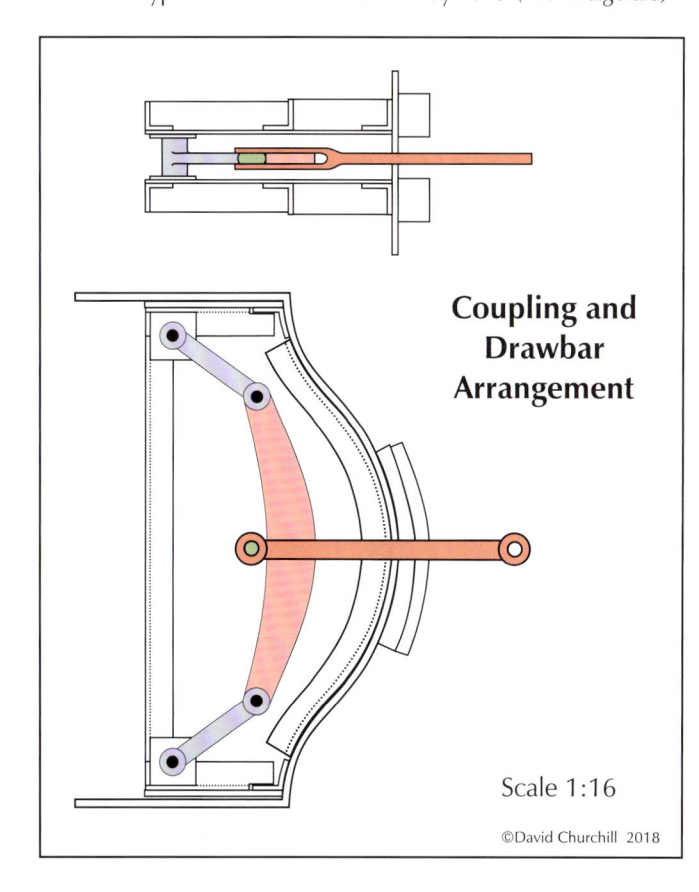

Coupling and Drawbar Arrangement

Scale 1:16

©David Churchill 2018

Couplings

It was recognised from the beginning that the tight radius and reverse curves would be likely to cause problems with couplings between locomotive and train.

The 'No1' class locomotives had 'link and pin' couplings mounted on long arms pivoted at approximately the centre of the locomotive. These were said to be particularly needed for the six-wheel Cleminson wagons used in the early days. The arrangement must have proved somewhat inconvenient as the arms ran under the ashpan.

With the introduction of the 'A' Class there was a change to a different and rather unusual arrangement, as shown on page 13, but still retaining the link and pin plus side chains.

Sharp Stewart's order book called it a 'Stradal' coupling. It was devised by R Stradal in Europe during the 1860s and was described as *'Improved Coupling for Locomotive and Tender for Traversing Sharp Curves'*. It had been used to a small extent on the Semmering & Brenner Railways and on some Avonside rack locomotives built for New Zealand in 1874.

In the *Railway Engineer*, Prestage called it the 'rudder coupling' and recorded that *'the rudder coupling was the outcome of some very careful experiments and tests to advance which we had the benefit of the experience of Mr Molesworth'* and that *'Its use has resulted in the life of the tyres of the leading wheels of the locomotives being increased from 15 months to some 4½ years'*.

Baldwin's description of it in 1916 was *'Drawgear to consist of coupling bars with pin holes at the end which are attached at the front and back of an engine to a swinging curved equaliser, which permits of great freedom in pulling wagons around curves of 60 feet radius'*. (Appendix 6).

Whatever its name, it proved very successful on the DHR and continues in use on the 'B' Class to the present day. A diagram of the coupling arrangement is shown on the drawing [left].

In about 1911, the DHR had a dalliance with a version of a coupling produced by ABC of Wolverhampton, which was quite widely used in India, for example on the Kalka-Simla railway. They were specified and fitted on bogie wagons, and apparently on the Pacific and Garratt locomotives. No photograph or other indication of it being fitted on a 'B' Class has so far been found, but it does appear that they could have been made compatible with the normal DHR couplings. However, whether fitted to a 'B' Class or not, they were abandoned in 1916, Mr Addis, the DHR General Manager, describing them as *'a bastard variety of the ABC. The breakages were very heavy'*. 'B' Class used elsewhere generally had the normal DHR arrangement, an exception in recent years being 794 at Matheran.

[Left] The 'rudder coupling' was introduced on the 'A' Class, proved successful, was continued on the 'B'Class and is stll in use today. The drawing shows the rear coupling on the 'B'class. A similar layout is used at the front although the dimensions differ.

The upper view shows a section on the centre line, the lower plan view is drawn with upper plate and angles removed to show more clearly the moving parts.

Lights and power supply

Lighting was by oil lamp in the early days, supplemented from the 1890s by a Wells light mounted on the cab roof. The Wells light (A C Wells & Co) was a proprietary blowlamp-style, oil-fuelled light introduced in the 1880s. The fuel was in a tank, housed behind the right hand side bunker on the 'B' Class, which was pressurised using a hand pump. It was connected via a pipe to the burner on the roof. The burner

Locomotive Headlamp

Sharp Stewart supplied an oil headlamp with the early locomotives and North British continued a similar design with the new locomotives of 1913. They remained in normal use until the advent of electric lights in the 1920s.

(Based on a Sharp Stewart drawing).

incorporated a heated vaporiser to vaporise the oil before it reached the nozzle. It produced a very bright light and was widely used on construction sites. The Wells light was successful on the DHR and although running in the dark was limited, it continued in use when required until electric lights were provided. A 1916 report of a journey says *'For the last 2,000ft the train runs through a dense jungle which completely shut out the fast failing light. A stop was made to allow the driver to light a huge flare on the roof of his cab – very similar to those used by coasters at home'*.

Electric lights on the locos were introduced in the 1920s powered by steam-driven turbo-generators from various makers, including Sunbeam. At first they were fitted at the right side rear of the bunker, immediately in front of the cab. Later they were moved down to the right side running board, perhaps to make life more comfortable for the coal breaker. From the 1950s the turbo-generator was moved to the left hand running board, with the exhaust taken across the front of the bunker and then vertically upwards.

Two views of the valve gear on 805 in 2006. Although the engine was in poor condition awaiting overhaul, the good light illustrates the layout. Note the saddle tank support/former sandbox without a front cover, and, on one view, the overflow from the injector fixed to the running board edge.

Sanding

Sanding is essential to the DHR because of the severe gradients and the often damp rail conditions. Effective methods of sanding have been sought since its opening.

When built, the 'No1' class had a single sand pot mounted on the boiler top with gravity feed to the front of the wheels when running forwards. Photographs suggest that this did not last long and that hand-sanding must have been used instead. For the 'A' Class, sand boxes were positioned on either side of the boiler above the valve gear, perhaps not an ideal position because of spillages during the frequent refilling. The sand feed was by gravity to the front of the leading wheels and controlled by a linkage from the cab.

The first four 'B' Class ordered in 1888 were fitted with steam sanding. This was added to the order in October 1888. At the time, steam sanding was a new development, having been

Mechanical lubricator fitted to No30 in 1914 *(DHRS archive)*

A locomotive at Tindharia around 1910. The handrails on the driver's side are filled in – which was sometimes done at that time. The controls are basically similar but simpler because there was no hydrostatic lubricator or turbo-generator. Note the helmet!

The Detroit No32 four-feed lubricator on 795 in 2015, although only two feeds are connected *(Nick Fitzgerald*

The Wakefield Eureka two-feed lubricator on 777 at Delhi *(Subhabrata Chattopadhyay*

introduced on the Midland Railway as recently as 1886. In theory it should have put down the sand just where it was needed and, being more controllable, would use less sand. However, it seems to have been a dismal failure on the DHR, as a later report noted *'experience having shown that steam and sand blasts were useless.'* It was abandoned for the second batch of 'B' Class locos, supplied just three years later: they reverted to gravity sanding operated via a linkage from the right side of the cab. This was retained for future new construction up to and including the final North British loco built in 1927 and those built for Raipur.

On the 'B' Class, the sandboxes were within the heavy brackets that supported the saddle tank from the frame. The gravity sanding was reported to be prone to blockage, difficult to control sensitively and hence wasteful of sand, and it seems that hand sanding was still in regular use. As an example, a

A lovely cab view of 786 being prepared for duty at Darjeeling in May 2015. It is full of interesting detail: part of a Ghum station sign has been used to repair the back of the bunker. The coal on the cab floor perhaps suggests why the driver almost always has a wooden platform to stand on. Tools are hung from the right-hand bunker-side rather than from the rerailing bar. The right-hand footstep support has been repaired using a flat plate. The wooden board maybe provides a more comfortable seat for the driver when running backwards *(Nick Fitzgerald)*.

By 1971, 783 had acquired the later angle section front roof support, although it was nominally one of the earlier locomotives. It also has a flat bar running down from the front of the cab roof to the coal rail. *(Michael Bishop)*

[Above] It is interesting that among the various modifications made to 19B during its rebuilding, a mechanical lubricator has been fitted on the right-hand side running plate. It is driven from the motion.

An enlargement from a 1960s picture shows this same metal strip [see far left]. This was often present in the 1960s and 1970s running from front of cab roof to bunker. The markings suggest that the bar was there to obtain a measurement of the volume of coal loaded *(Lou Johnson)*.

[Above] This view of 785 in 1982 shows clearly the modified rear cab roof profile and also provides a view of the interior of a completely empty bunker *(Chris Pietruski)*.

[Left] Having already left India, 19B escaped the cab roof modification and retained a rectangular roof. Though damaged, it is seen shortly after arrival at Tyseley in 2003

description of a journey made in 1915 records *'Instead two natives sit on the buffer beam and, leaning forward, sprinkle the rails with sand as the train goes along.'* The familiar DHR image of men sitting at the front of the locomotive ready to apply sand to the rails was clearly already well established.

Eventually, by the 1930s, gravity sanding was given up altogether, the operating linkage from the cab removed and hand-sanding alone relied upon, with the sand supply carried in the toolbox at the front. The sand filler was in many cases removed from the front of the former sandboxes, which became toolboxes. Some of the locos do still have the sandbox type lid today, for example 785, 786 and 793. When done skilfully, hand sanding was more efficient and economical in use of sand. However, it is often still necessary to stop en-route to top up the sand supply. To make life easier for the sanders, handrails on either side at the front were provided by the 1940s.

The issue of sanding was addressed again in the 1960s, possibly in an attempt to reduce staff costs by eliminating the sanders. 790 was briefly fitted with very large sandboxes on either side of the saddle tank. How successful they were is not known but they had been removed by the early 1970s. It is illustrated on page 99.

Thereafter, hand sanding has remained in use on the 'B' Class to the present day and the Golden Rock new-build and refurbished locos still use it. The NDM6 diesels introduced in 2000 have four sandboxes per loco, with a total capacity of 100 litres, allowing sanding when running in either direction. The system appears to work, with no sign of hand-sanding having to be used. 19B is now fitted with air-operated sanding for both forward and reverse running.

Wooden re-railing bar.

Until recent years a wooden re-railing bar was normally carried on the right-hand side of the loco, supported on brackets at the back of the bunker and front of the saddle tank. Similar brackets were provided on the left side but it was very unusual for a second bar to be carried there. The only occasions this was noted being for the Cinerama film in the 1950s and on the loco prepared for the visit of the Prince of Wales in 1904.

When the bunker side height was raised by adding coal rails during the 1950s, a cut-out had to be provided on the right side for the beam – this was occasionally provided on the left side as well. Since the early 2000s, carrying the beam seems to have largely been discontinued. The refurbished locos have no brackets or cut-out. The locos supplied to Raipur carried the beam further out, with the front bracket on both sides of the saddle tank angled outwards to suit. This feature remained on the ex-Raipur locos for years, but it has eventually become spread amongst various survivors.

19B in 2017 still retains its original design of cab front that remained on some locos long after they were fitted with dome-mounted safety valves. It shows clearly here, together with the extra section added to raise the cab roof when it as rebuilt at Tyseley. It also shows Adrian Shooter's alternative narrower cab roof fitted for clearance when operating on the Ffestiniog Railway. Note also the pillar supports for the front of the cab roof.

780 immaculate at Siliguri shed in February 2006. Note the electric lights which illuminate the gauge glasses with a connection to the electric supply. The driver's wooden platform protects his feet from the coal usually on the footplate.

In 1980, 790 had a Belpaire firebox. The speedometer housing is still present at the top left of the cab, although the dial and linkage have been removed. An oil lamp is in place to illuminate the left-hand gauge glass. The extension platform is in place at the rear for the fireman when running in reverse up to Ghum.

92

792 is captured in the evening sun in 1979, when it had the usual bunker extension plus additional coal rails.

Bunker extension

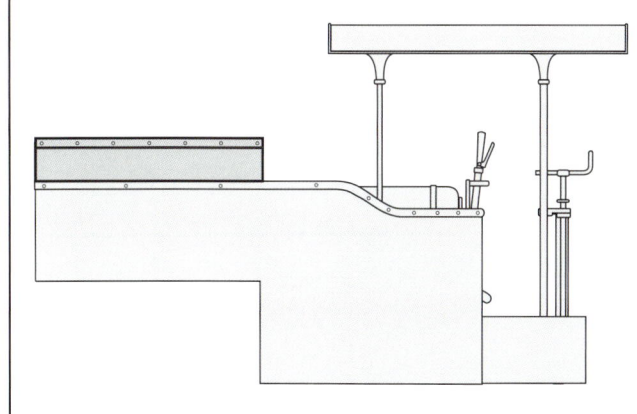

The bunker extensions (coal rails) to increase the coal capacity were fitted to all the 'B' Class on the DHR during the 1950s. The pattern varied somewhat and pictures of many examples can be found in this book.

The drawing above is based on one dated 1954 which showed a style which does not extend back to the cab.

The bunker height was extended by 9in on all four sides with 2 in x ⅛ in flat bar around the top as beading.

No photograph of this arrangement has so far been seen so it may never have been implemented.

©David Churchill 2018

A view of the inside of the bunker of 19B at Leighton Buzzard in 2013. It is no longer used for coal but does illustrate the construction and the two rear supports for re-railing bars still in place. 19B uses air-operated sanding and the two sandboxes are can be seen on either side of the bunker.

Two pictures of a locomotive in Tindharia works in February 2006. The insides of the bunker extensions are rusty sheet metal, supported on angles fixed to the bunker sides. The picture on the left shows the pillar that supports the Detroit lubricator. That on the right shows the reversing lever and the handle that operates the cylinder drain cocks.

[Far left] A headlight from the 1930s and 40s. Often, as here, they carried the locomotive number.

[Centre left] An example in use in the 1970s *(Michael Bishop)*

[Left] The headlight currently fitted to *777* in Delhi Museum.

The former sandboxes still prove useful where they survive. Presumably the bottles contain oil. 786 is at Darjeeling 2015 *(Nick Fitzgerald)*.

Speedometer

Train speeds and speed limits were a concern on the DHR from the earliest days. In 1881 Franklin Prestage assured the Government Inspector that speed-indicating equipment was *'on its way from England'*. There was no further mention of it; whether it never arrived or failed to work is unknown.

During the 1960s there was another attempt at fitting speedometers. They were mounted on a bracket on the left side of the loco and driven from the eccentric. A cable linked it to the display in the top left corner of the cab. The readout equipment appears to have been supplied by VDO. Use of the speedometers was short-lived, although remnants of the brackets remained on the footplate of some locos for years and are indeed still present on some of the plinthed locos.

In 1981, an interesting test followed the derailment of 10D Down Passenger between Darjeeling and Ghum. The maximum permissible speed at the time for a down train between Darjeeling and Sukna was 16km/h. The investigators concluded that excess speed was a contributory factor and their report stated that *'In the absence of speedometers, drivers are not in a position to judge speeds accurately'* and also that *'there appears to be no serious effort...to monitor speeds and to counsel or discipline drivers needing such treatment'*.

A trial run was organised with 806 and four coaches from Darjeeling to the site of the accident and the driver instructed to drive as usual. The maximum speed approaching the site of the accident was calculated as 19.44km/h. An up-to-date feature on 19B is its GPS speedometer!

DHRS Engineering Project

In 2011-12, the DHRS supported an 'Engineering Project' in which David Mead and Mike Weedon visited the DHR. The part of the project relevant here was overseeing the overhaul of a 'B' Class at Tindharia.

The condition of the locomotives had been gradually deteriorating, with train loads reduced to two bogie coaches from the four that had been possible at one time. The engine chosen was 806,

It is necessary to stop to replenish the supply for the sanders occasionally. 804 is seen above Rangtong in 2006. Note the locomotive has a 'blank' cylinder cover without a connection for a tail rod. Also, as an ex-Raipur locomotive, it still has a splayed-out rerailing bar support, although it is rather bent!

Mike Weedon is seen using the purpose-designed, in-situ cylinder boring machine which was available at Tindharia to re-bore the cylinders *(David Mead)*.

which was being used as Siliguri shed pilot because it was unable to haul a loaded train up the line.

It was evident that much steam was blowing past the pistons, and after it was taken to Tindharia, removing the pistons revealed badly-worn cylinder bores, under-size pistons and loose piston rings. The cylinders were rebored, new piston heads and rings were made and on a later test run 806 hauled three coaches with ease, although other troubles (including difficulty with the injectors) prevented a run with four coaches.

[Right] The original piston heads had one wide groove with two piston rings side by side in the single groove. This was an early Victorian design prone to leakage. New piston heads were made by Mike Weedon and fitted to 806 with two separate grooves with a single piston ring in each groove. The results of this simple improvement together with the cylinder rebore astonished the railway's staff *(Mike Weedon / David Mead)*

[Below] David and Mike were also asked to help with the valve adjustment on 791 at Darjeeling. They fitted a turnbuckle adjuster to the union link, to make adjustment easier than the normal method of shortening or lengthening of various links. The modified link is seen here with the original union link on the floor underneath. The valve adjustment had been a long way out and the result was dramatic – the Joy Train arrivied at Ghum 10 minutes early! *(David Mead)*

96

Brakes

The early days

Braking of DHR trains was a major concern even before the line opened and the 'No1' Class locomotives (and later the 'No2' Class) were supplied with three completely independent brakes:

1. A conventional handbrake acting on the wheels, operated from a vertical brake column in the cab
2. A skid brake acting directly on the rails operated through a completely separate linkage from another brake column in the cab. Sharp Stewart had previous experience of this arrangement, having used it some years earlier on some rather unsuccessful broad-gauge engines for the Bhore Ghat near Bombay.
3. A Le Chatelier-type counter pressure brake. This seems to have been considered as a system of last resort and is not often mentioned – Prestage in 1887 said it was rarely used. The details of its application to the DHR engines are not clear. Nothing obvious is visible in early photographs.

The issue of braking was considered so important that during his pre-opening inspection of the first section of the DHR on August 14 1880, the Government Inspector arranged some *'rough practical trials – speed judged by eye only and that the full train tests were made under exceptionally trying conditions during a severe storm'*. Some of the results are summarised below. The Le Chatelier brake was not mentioned in the report.

DESCRIPTION OF VEHICLE OR TRAIN	BRAKE POWER	GROSS WEIGHT VEHICLE OR TRAIN	GRADIENT IN	SPEED MPH	STOPPING DISTANCE (FT)
No1 locomotive	Wheel brakes	10 tons	22	7	17
			22	10	25 / 30
			22	14	31
			22	20	65 / 62
	Skid brakes		22	10	52 / 50 / 80
Full train	All wheels braked	20 tons	23	7	35
			23	10	100
			23	12	120
	Only engine wheels		23	7	65 / 75
			23	10	100

This was obviously considered acceptable, but concerns remained. In 1882 and 1884, there were further tests with similar results. However, in his report for the half year ending 31 December 1889, Lt-Colonel H.Wilberforce Clarke dismissed the earlier tests saying: *'These experiments are (I regret to say) nearly valueless…'*.

Skid Brakes

Routine use of the skid brake was creating problems and additional Subsidiary Rules were drafted in 1887 to try to stop its use except in emergencies. These included:

8. *Before any train leaves Ghoom for Darjeeling, or for Sookna, or any intermediate stations down-hill, the Driver of the train will see that the jemadar brakesman releases the levers of all brakes, and if necessary pins them down so as to put on the pressure required.*

9. *The Driver must see that sufficient pressure is put on the brakes of the vehicles forming the train to hold the train to a speed of 7 miles an hour, as ordinarily the brake-power exerted on the engine should only be sufficient to hold it alone, the engine wheel brakes only being used.*

10. *These instructions, which should be strictly observed, are issued mainly to prevent the constant and free use of the Skid brake excepting in cases of emergency.*

The first three orders for 'B' Class locos retained the skid brake for emergency use, but dispensed with the Le Chatelier brake.

The arrangement of the skid brake on the 'B' Class and its operating linkage is shown on page 98. The locomotives still had two separate brake shafts with two vertical handbrake columns, one linked to the normal wheel brake and the other to the skid via completely separate linkages.

However, the trains and locomotives were now considerably heavier and in May 1893 S B Cary, the DHR General Manager, proposed some official tests to determine its effectiveness. Lt-Colonel Little's Inspection Report for the year ending December 1892 described these (the inspection was very late!) as follows: *At the suggestion of the General Manager, some experiments were made in a length of straight line near Tindharia to test the value or otherwise of the skid brakes, two of which are on each engine, in addition to the cast-iron blocks one to each wheel, all worked by hand levers.*

The gradient of the length selected to make the trials on was 1 in 27 and the line was practically straight. The weight of the engine was 14 tons and the train load was 700 maunds or 25.7 tons carried on 7 wagons, each weighing approximately one ton, making the total weight of engine and train 46.7 tons. The handle for the skid is close to that for the ordinary wheel brake, so that it is impossible to work both together efficiently or rapidly when the train is pulled up in a short length as it was in trials 1, 2, 6, and 7.

In trial No.3 with the skid only, the driver, after going 340 feet and finding that the speed was increasing dangerously, had to reverse his engine to bring it to a stop.

TRIAL NO.	NUMBER OF VEHICLES	GROSS LOAD	SPEED	DISTANCE IN WHICH STOPPED	REMARKS
1	Light Engine	14 tons	12 miles	60 feet	Ordinary brake only.
2	Ditto	Ditto.	Ditto.	60 feet	Ordinary brake and skids.
3	Ditto	Ditto.	Ditto.	Not stopped	Skid only.
4	Engine and seven trucks	46 tons.	Ditto.	252 feet	Ordinary brake on engine only none on wagons.
5	Ditto.	Ditto.	Ditto.	240 feet	Ordinary brake and skids on engine only, none on wagons.
6	Ditto.	Ditto.	Ditto.	60 feet	Ordinary brake on engine and all wagons braked.
7	Ditto.	Ditto.	Ditto.	60 feet,,	Ordinary brake on engine and skids and all wagons braked

From this it appears that the skids re for all practical purposes useless, and it is quite within the bounds of possibility, I think, that they be a source of danger, as if they were put on very tight they might take weight off the wheels and thus reduce the action of the hand brake which really is the only effectual means of stopping short of reversing the engine.

It seems that Mr Cary was not comfortable with totally abandoning the skid brake on his own initiative and it was 30 June 1896 before he wrote to the Senior Government Inspector of Railways to 'solicit formal sanction to remove the skid brake from our engines'. The reply of 24 July confirmed that the Government of India had *'no objection to removal of the skid brakes now attached to the locomotive engines in use on the Darjeeling Himalayan Railway'*.

The report of an inspection in April 1897 confirmed that the skid brake had by then been removed from the 'B' Class and earlier locos that had it fitted. The skid brake was not included in any future 'B' Class orders.

Surprisingly, a remnant of the skid brake seems to have survived on 19B. Although no longer used, the bearings for the second forward brake shaft are still present.

Hand Brake

Braking was now just the conventional hand brake acting on the engine wheels only, plus an appropriate number of brakesmen on the vehicles of the train. The DHR's safety

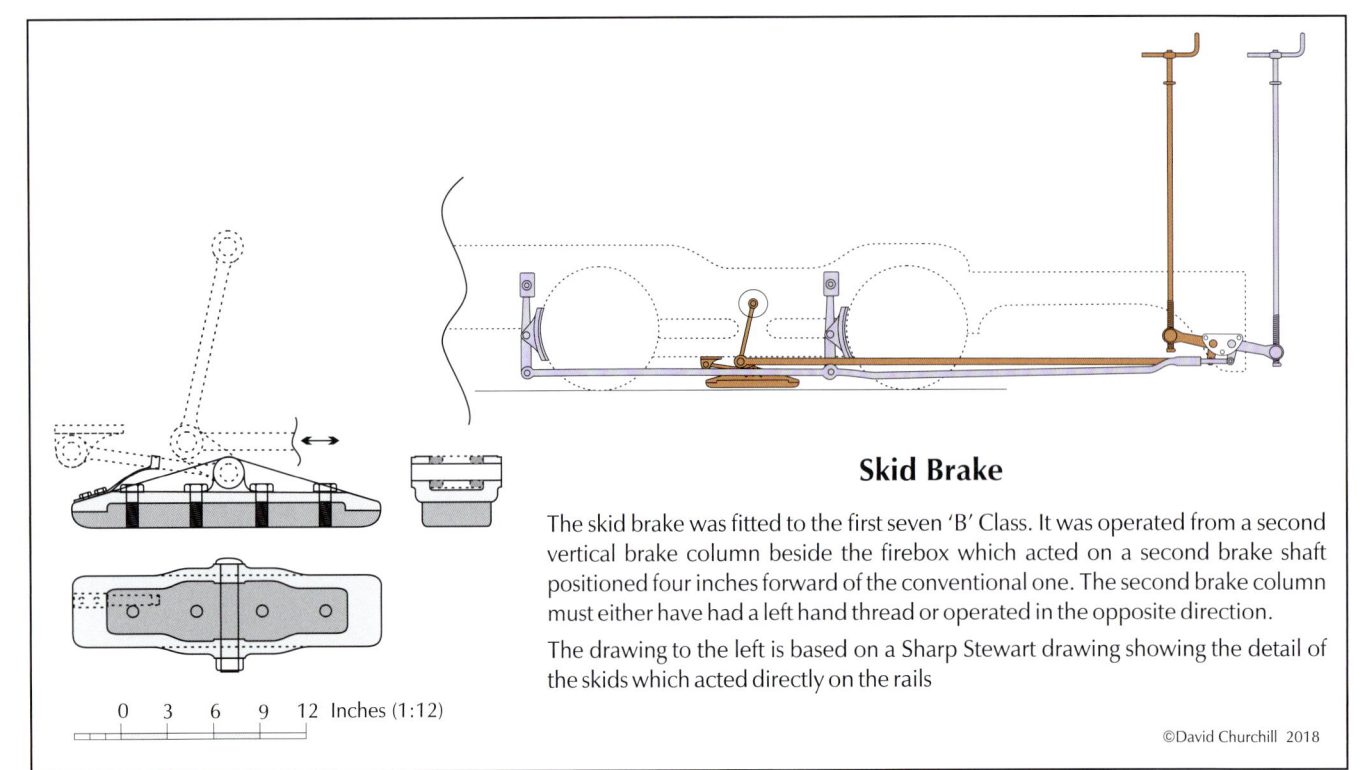

Remnants of the skid brake brake shaft on No19B – September 2017. The bearing plate for two brake shafts can be seen still in place on right-hand side side of the loco. The cylinder and link at the front is part of the air braking now fitted.

Skid Brake

The skid brake was fitted to the first seven 'B' Class. It was operated from a second vertical brake column beside the firebox which acted on a second brake shaft positioned four inches forward of the conventional one. The second brake column must either have had a left hand thread or operated in the opposite direction.

The drawing to the left is based on a Sharp Stewart drawing showing the detail of the skids which acted directly on the rails

©David Churchill 2018

record was good but the Government Inspectors were still concerned, for example in 1905 and 1906, Mr J H White wrote: *'Constant attention to the efficient state of the brakes on every engine and car and frequent tests of the same are absolutely necessary for safety. The responsibility for maintaining the brake power up the standard and the efficient manning of the same rests with the Administration. The necessity for vigilance cannot be too strongly impressed on the train staff and the trains should be practically tested periodically. Where the application of the brakes cannot avert danger, the safety of the train depends upon the alertness of the driver and brakesmen, the brake power is sufficient provided the speed is not excessive'.*

The DHR standing orders included. *'1.Engine Brakes to be daily examined on the Engines coming in off their runs and again before leaving the shed.'*

Since 1896 the 'B' Class have continued to rely on hand brakes alone, plus the brakesmen on the train, again with a remarkably good accident record. The line's later Garratt and Walford diesel engines and several unbuilt proposals all included some form of power braking. Some coaches were air-braked in the 1940s to operate with the unsuccessful diesel but no evidence has yet been found of any attempts to fit a 'B' Class for air braking at that time. There were later trials with both vacuum and air brake but both were short-lived, as described below.

Vacuum brakes

In the 1950s and 1960s efforts were made to fit 'B' Class locos with vacuum brakes. These were perhaps prompted by concerns raised when the DHR became part of the Northeast Frontier Railway. Very few records have been found of these trials so what follows is based on study of photographs and a few somewhat confusing and not very clear drawings. Fitting a vacuum brake cylinder on to a 'B' Class is not easy as, unless an additional linkage of some sort is used, it needs to be behind the brakeshaft to act in the correct direction.

Photographic evidence shows that at least one 'B' Class, No 35, was fitted with vacuum brake equipment as early as the late 1950s or early 1960s. The photograph shows:

1. A vacuum pipe connection at the front right-hand side of the loco
2. A reservoir under the left-hand side of the footplate – a drawing exists with a reservoir in this position
3. A vertical pipe in front of the cab, assumed to be the ejector exhaust
4. A slotted link at the bottom of the handbrake screw – necessary if the loco is to have both the power brake and handbrake acting on the same shaft.

No vacuum cylinder is visible but it is presumed to have been squeezed under the right side rear of the footplate, necessitating a modified brake shaft arrangement and altered footstep. This presumption is supported by the same loco, by then numbered 790, retaining well into the 1970s its unique arrangement of a different right-hand cab footstep

A picture dating from late 1950 or early 1960s. Loco No 35 fitted with vacuum brake equipment. Note the swan-neck pipe for the vacuum pipe connection on the right-hand side front of loco. No flexible connection is fitted – in fact it seems to have been appropriated to fit to the end of the Turbo-generator exhaust. The vacuum reservoir can just be seen under the left-hand side of the cab. The vertical pipe in front of cab would be the vacuum ejector exhaust. There is a fitting on the top which directs it to the rear – perhaps to benefit the coal breaker.

The same loco has now been renumbered 790, c1975. This picture shows clearly the horizontal link for a vacuum cylinder, also the forward brakeshaft with a hole where the original brakeshaft had been and the different footstep that had previously also supported the vacuum cylinder.

A few years earlier, 790 is pictured in 1970. Much more of the brake equipment is in place – the train pipe at the front, the reservoir on the cab roof and the forward brake shaft. The large sandboxes are still in place, together with speedometer *(Laurie Marshall).*

99

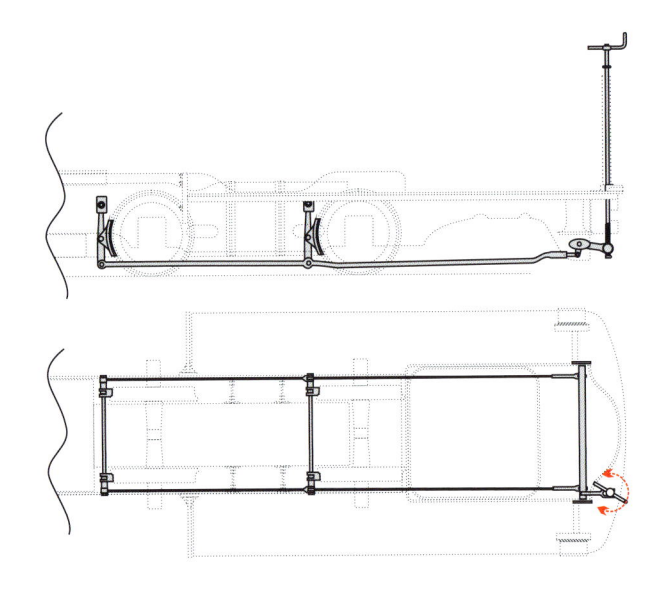

Handbrake operation

The diagrams show the elevation and plan (viewed downwards) of the standard hand brake operating linkage. This was the only brake on the locomotive after removal of the skid brake in the 1890s and has remained basically unchanged to the present.

1960s Vacuum and steam brakes

Ⓐ The modified footstep to accommodate the extended brake shaft.
Ⓑ The slotted link added to the linkage at the base of brake column.
Ⓒ Details of linkage uncertain.

The diagrams illustrate the brake arrangements for the 1950s and 1960s vacuum brake trials. They are based on a study of photographs and some unclear drawings and some uncertainty therefore remains.

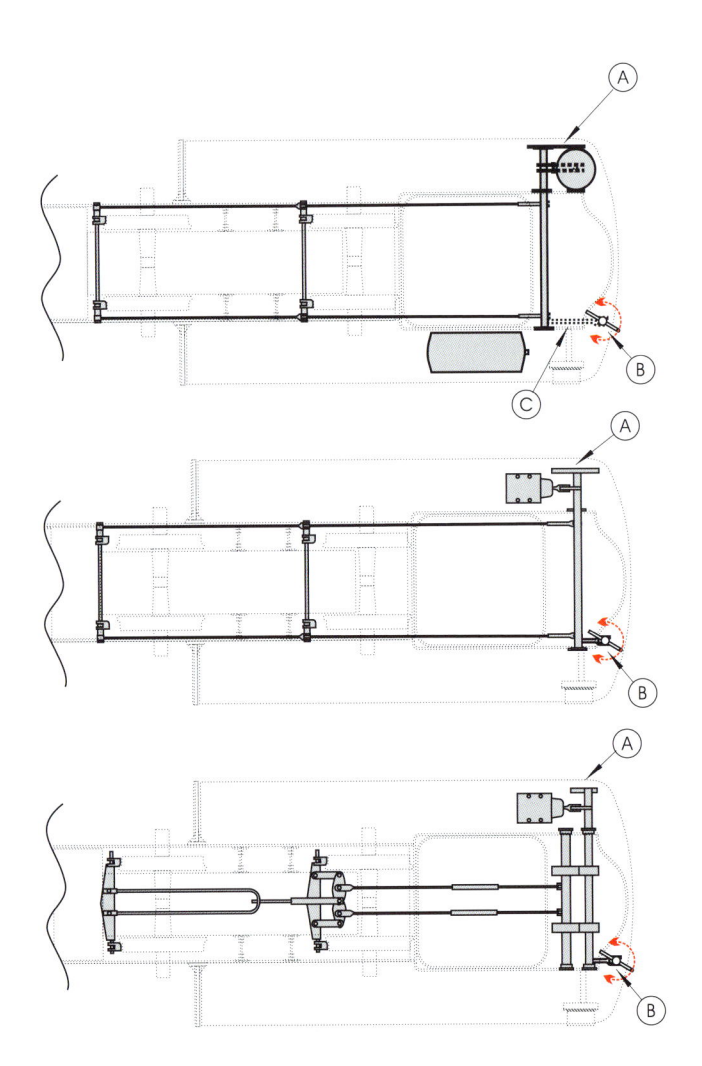

790 was fitted with a vacuum brake on the locomotive. The normal brake shaft was removed and a new one fitted close behind the ashpan. The vacuum cylinder was fitted behind a modified right-hand footstep which also provided a support for it. Details of the linkage between handbrake column and new brake shaft are uncertain. The vacuum reservoir was first under the left hand footplate, but later moved to cab roof.

In the 1960s, a number of locomotives were equipped for a vacuum-braked train with a steam brake on the engine. The diagram illustrates the arrangement used. This necessitated a rear brake-shaft extended on the right side and supported by a modified footstep.

A drawing exists with, handwritten, the number 798. It shows a peculiar system for a steam brake on the engine with twin brake shafts linked by a pair of gears. It also shows an equalised brake linkage rather than the normal pattern. No evidence has been found that this arrangement was ever implemented. However 777 still has unused connections which appear to be for an equalised linkage on its otherwise normal brake system.

©David Churchill 2018

803 pictured is in 1971. The train pipe runs along the right-hand side of the locomotive: the reservoir is on the cab roof *(Ian Fraser)*

804, now plinthed at New Bongaigaon, has an unusual hand-brake column. Pictures show that around 1970 this column, which seems to be unique, was on 798 that is known to have been involved in the vacuum brake experiments, and in fact its number is pencilled on one of the drawings. Could it be a column arranged to push rather than pull?

This picture also illustrates another question – the rear right-hand end of the angle is cut away. This is not a random thing: there were several that were similar. Was it connected with the aim of getting clearance to hang the vacuum cylinder from the back of the footplate, where it would have been behind the normal brake shaft? *(Nick Fitzgerald)*.

Much of the vacuum brake equipment, including the brake cylinder, has been removed on 804, probably in the 1970s. However, the extended rear brakeshaft with, in this case, a vertical arm to connect to the brake cylinder is clear. The slotted link at the bottom of the handbrake screw can also be seen.

777 in Delhi Museum still has the extended rear brake shaft and modified footstep.

A drawing exists showing an 'equalised' brake linkage inside the wheels, rather than the normal arrangement outside the wheels. This was shown roughly on a contemporary drawing marked experimental. No evidence has yet been found of a loco running with this in service, but 777, which has been in the Delhi Museum since it opened, has the connections on the rear brakeshaft that would suit such a linkage *(Mick Melbourne)*.

101

and forward-mounted brake shaft. It appears to have been part of the late 1960s brake trials and is a particularly interesting loco as, in addition to its unique brake system, it was the loco fitted with enormous sandboxes.

In the late 1960s, there was another major effort to update the DHR brakes – the new coaches delivered in 1967/8 were all fitted with vacuum braking, as were some of the four-wheeled vans. Further trials were conducted to adapt the 'B' Class to run with them. 790, discussed above, was one example, and Nos 798, 802, 803 and 804, at least, were fitted with most, if not all, of the fittings needed for a rather different arrangement. There appear to have been experiments, or at least designs, for several system variations.

One system appears to combine a steam brake on the loco with vacuum braking on the train. This had a steam brake cylinder tucked under the right-hand footplate of the loco, acting on a vertical arm on the normal brakeshaft extended at its right-hand side to a bearing in a modified footstep. The link at the bottom of the handbrake screw was slotted as before with a modified brake linkage and the vacuum ejector fitted in the cab. The train pipe wound its way, rather untidily, along the right side of the loco. There was a reservoir on the cab roof. The ejector exhaust was taken through the cab front and vertically upwards, except on 798 where it was taken forward alongside the train pipe and into the smokebox.

There are several puzzles remaining: nothing on the DHR is ever straightforward. These are illustrated by the photographs.

The vacuum brake trials ceased around 1970. The reason for their failure is not known for certain. It was possibly a combination of technical issues (for example, steam consumption) and the fact that hoped-for staff savings were not realised.

Air brakes

By the early 2000s, air-braked coaches were available on the DHR, and after early teething troubles the NDM-6 diesels were able to operate with them. This led to a desire to fit the steam locomotives with air-brake equipment, and it was included in the new-build 'B' Class specification.

The trial oil-fired engine 787 was modified with an extended brakeshaft on the left-hand side, with its brake cylinder mounted under the footplate and appropriate new linkage. New loco 1002 was similarly fitted. However, it seems that these air brakes were never used in earnest. When 1001 eventually arrived at Siliguri in January 2007, it had no air brakes, although they appear to have been fitted earlier in its life at Golden Rock.

In 2008/9 some further air brake trials were carried out using 'B' Class 804. On this occasion the air supply came from a diesel-driven air compressor installed in a support coach, with brake valve and other control equipment on the locomotive itself. Twin pipe connections were fitted at the front and rear, with external pipework along the right-hand side. No air brake was fitted on the loco itself. The trials were reported as being successful, but development was not pursued any further and the equipment was soon removed.

In March 2018, following the Chairman of the Railway Board's visit to the DHR, is was announced that all 14 'B' Class locos still on the DHR were to be fitted with air brakes.

804 in the NG platform at New Jalpaiguri station in January 2009 with most of the air brake equipment still installed but not in use *(Keith Froom*

The driver's valve on 804 *(Ian Fraser).*

Liveries, Names and Plates

Green

From the beginning through to the 1940s, the locos were green. There is little direct evidence of the colour used – no samples are known to survive- but, if the modern blue colour is any indication, it is likely to have varied considerably over the years.

One description had the colour as being 'like Great Northern green'. This seems quite specific and GNR Society members have advised that Phoenix Paints P750 and P751 are judged to be a good match to two versions of GNR green.

The Baldwin specification of 1917 states *'ship with locomotive a sufficient supply of Napier Green enamel to be used for painting locomotives after their erection. Herewith small sample of Napier Green enamel'*. The sample has unfortunately has not survived. The term 'Napier Green' seems to have been derived from the colour used for Napier cars in the 1900s – a predecessor of British Racing Green, but somewhat lighter. It appears that when Napiers moved to Acton, the doors and window frames of the works were painted with a shade of olive green supplied by Carsons, a London paint manufacturer. Apparently the same paint sufficed for both factory buildings and racing cars. Walter Carson & Sons were very long established suppliers to both railway companies and the Indian Government, and advertised in the railway press. Therefore it seems very likely to have been paint suppliers to the Darjeeling Himalayan Railway. It seems logical to conclude that Carson & Sons would have adopted the term Napier Green for one of their colours after its use by Napier on their cars and factory buildings.

A report from a visitor to the DHR in 1945 also describes the loco livery as 'olive green, lined black and yellow, brass fittings burnished'.

806 in its green livery at Kurseong station whilst shunting a special train on 28 April 1995 (Peter Jordan)

A still captured from a 1940s home movie of Batasia loop. Not good quality, but it is the only colour image yet found of one of the grey locomotives (Cambridge University Cen. South Asian Studies)

777 with its accompanying DHR coach pictured soon after opening of the Delhi museum in 1977. When prepared for the museum the locomotive was painted in a green that was believed to be a good representation of the DHR loco colour. It was lined in black and yellow as used from the 1920s. The IR numberplate would, of course, not have been carried at that time (George Woods)

In the early years, locomotive numbers were painted in large shaded lettering on the bunker side as NoXX. From about 1920 this was replaced by the number in cut-out brass figures on the cabside, below the worksplate, where one was fitted. At the same time DHR was painted on the bunker side in yellow shaded letters.

Lining on the engines appears at first to have been black, edged with red but later changed to black, edged with yellow. The date of the change is not definitely known but it is suggested it was about 1920 when the DHR lettering replaced the painted number on the bunker side.

Silver grey

During the Second World War several 'B' Class engines were painted in a light grey livery. The colour has been

103

described as 'light blue', but it is thought this is likely to be another description of the same colour. The only colour pictures found are stills captured from a home movie. These are not good quality but do confirm that the colour was very light.

Based on a study of pictures on page 38 and elsewhere the colours were:

- *Bunker and cab sides, bunker front, saddle tank including its front and back, front sandbox, dome cover: light grey*
- *Cab front: either black or dark grey*
- *Boiler: black*
- *Frames: Polished metal*

Lining appears to be a single wide line on the cab, bunker sides and saddle tank (not on the tank front and back). The lining colour was probably dark grey.

Black

The locomotives are recorded as being painted black for a time at the end of the Second World War. There is no evidence that this lasted for long on the 'B' Class, although the DHR Pacifics remained in black into the 1960s.

Red

The North Eastern Railway took over the DHR in 1952 and soon introduced a new red unlined livery. The colour has been described as 'maroon', as 'terracotta brick red' and as 'iron red'. Photographs suggest that it varied quite considerably.

The locos were lettered 'NE' in yellow-shaded black and continued to carry the DHR brass numbers on the cabside, usually below the builders' plate. The regular crews' names were often painted on the cabsides.

A later variation lost the brass numbers and crew names and the NE lettering was painted in yellow inside an ellipse on the cabside, with the loco number and class in a circle. This scheme was carried on into the NF railway era, with painted 'NF' replacing NE.

The buffers were generally red, and a much brighter shade than the other paintwork. With the All India renumbering of 1957, standard plates were gradually introduced carrying the Indian Railways numbers. The background of the plates was usually black on red locos. At the same time, brass lettering on the bunker sides appeared, replacing the painted letters. Some locos had two white lines around the bottom of the cab that swept upwards to run along the bottom of the bunker,

The red livery lasted until the mid-1960s when the now familiar blue was introduced.

Blue

Blue is reported as having been first applied to one of the locos in 1964 and has continued to be used, with very few exceptions, ever since. However the colour of 'Darjeeling blue' has varied considerably from a light blue to a dark 'Oxford' blue and it also varies considerably with weathering and cleanliness. At least one loco in the 1970s was so dark that it might actually have been black!

Builder's and rebuild plates

No attempt has been made to reproduce the lettering style used on the plates above. Refer to the photographs.

Original Sharp Stewart
DHR No 17 to No 23

Tindharia built locos
DHR No 42 to No 44

Later Sharp Stewart
DHR No 24 to No 28

Later North British
DHR No 45 to No 53

First North British
DHR No 29, No 30,
No 32 to No 36

Early Tindharia rebuild plate
also carried by 'A' Class

Baldwin
DHR No 39 to No 41
carried on smokebox

Later Tindharia rebuild plate
There was also a somewhat larger later version

Indian Railways numberplates lettering and nameplates

The NF lettering is in brass, normally in English on the left side of the locomotive and Hindi on the right.
All numberplates are similar and with similar lettering although the numbers are not always well lined up. The nameplate shown is an example of those introduced in the 1970s.

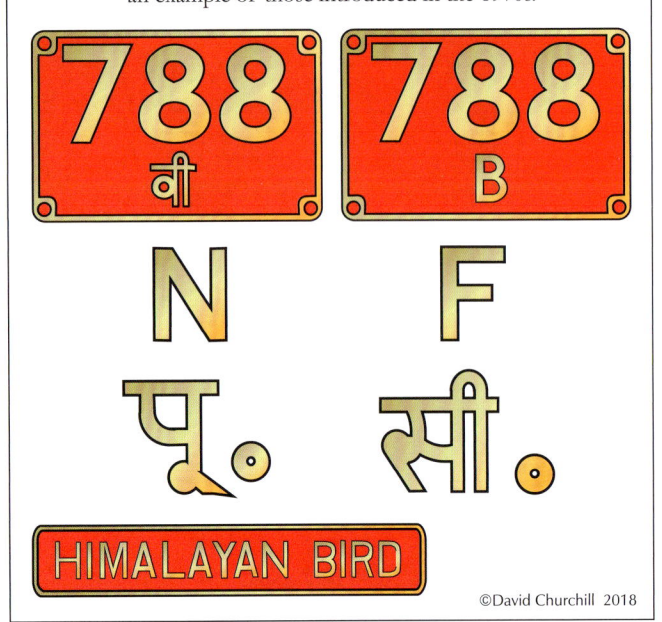

HIMALAYAN BIRD

©David Churchill 2018

There have usually been two white stripes around the bottom of the cab. At first these swept upwards to run along the bottom of the bunker on locos allocated to the Mail working.

The buffers have generally been painted red, and the footplate edges and motion bracket have often also been red. The background of the brass numberplate has normally been red on blue locomotives.

Short-lived 1990s colours

Although there have been lots of variations in the blue livery since it was introduced in the 1960s, there have been remarkably few cases of 'B' Class locos on the DHR being painted in any other colour. Two example in the1990s were 806 being liveried green for a time in the 1990s in a scheme similar to that used in the 1930s, although the green does appear to have been somewhat lighter. In 1997, 794 was very briefly painted in a red/yellow 'Shatabdi' colour scheme at Darjeeling.

794 at Darjeeling in its very short-lived 1997 red/yellow 'Shatabdi' colour scheme. It was reported that on seeing it, NFR officials ordered that it be returned to the usual blue *(Peter Tiller)*.

Number and builders' plates

The 1920s North British-built 'B' Class delivered to the DHR carried elliptical numberplates on their chimneys for many years. The latest seen on a photograph is on No 45 in 1959.

With the All India renumbering of 1957, standard numberplates were introduced. The plates had the number above English letter 'B' on the loco's left side and Hindi equivalent on the right. Very occasionally the plates were fitted on the 'wrong' side of the locomotive.

Names

All names were in English letters on both sides of the locomotive, except where indicated below.

The three Tindharia-built locos were named TINDHARIA (No 42), KURSEONG (No 43) and DARJEELING (No 44) – painted in an arc around the worksplate. It is not clear how long these names survived – quite possibly until the state takeover.

The streamliner carried the painted name JERVIS BAY for a time in the 1940s.

It was a surprise on our first visit to the DHR to find four 'B' Class engines carrying names on neat brass plates. It seems that they were introduced in about 1976. Except for GREEN HILLS, the names continue to the present day, usually on the same locomotive, although this has changed occasionally. GREEN HILLS does not seem to have been used on a working locomotive since 780 went to Golden Rock for rebuilding in 2009. However, it re-appeared on a plinthed locomotive, 780(M), at Kozhikode in 2013 (see page 71). The table gives some indication of names, loco numbers and dates, although it is not claimed to be complete.

	1979	1999	2006	2016
MOUNTAINEER	779	782	782	786
QUEEN OF THE HILLS	780	804	804	804
GREEN HILLS	798	780	780	......
HIMALAYAN BIRD	803	779	779	779

Around 2003, painted names started to appear on the 'B' Class locos. Names known to have been carried are shown in the table with approximate dates.

NAME	LOCO NO	APPROXIMATE DATES
WANDERER	780	2006 to 2015 (and see above)
AJAX	786	2004 to 2014
TUSKAR/TUSKER	788	2003 to 2014 (see page 62)
HAWKEYE	792	2003 to 2016
VICTOR	802	2004 to 2016
VALIANT	804	2006 to 2015
SOLDAT	805	2006 to 2015
QUEEN OF HILLS	806	2003 to 2015 but not continuously

It has been recorded that JUMBO (786), HORATIO (791) and BRONCO (795) were to be used, but no photographs have been found showing these actually painted on the

---

There are few works or rebuild plates still carried on locos. A number are now displayed in the Ghum museum, including the much-polished plate from Baldwin 44913 *(Peter Tiller)*.

locomtives. These painted names eventually disappeared as locos were repainted.

In 2005/6 the new-build GOC oil-fired locos were named as follows:

| 1001 | HIMRATHI |
| 1002 | HIMANAND |

These names were carried on rectangular plates, in English on the locomotives' left-hand side and Hindi on the right. The plates were soon removed from 1001, and 1002 was returned to Golden Rock.

In 2013, the former oil-fired 1001 was refurbished at Tindharia, renumbered as 01 and named TINDHARIA, the name being carried on a neat brass plate. In 2016, following a competition run locally, 802 was named WHISTLE QUEEN, also on a brass plate. In April 2016, 805 was given a painted name, IRON SHERPA. These recent names are illustrated in pages 63 to 73.

The 'B' Class, 794 at Matheran was noted as carrying the name MATHERAN QUEEN in 2006.

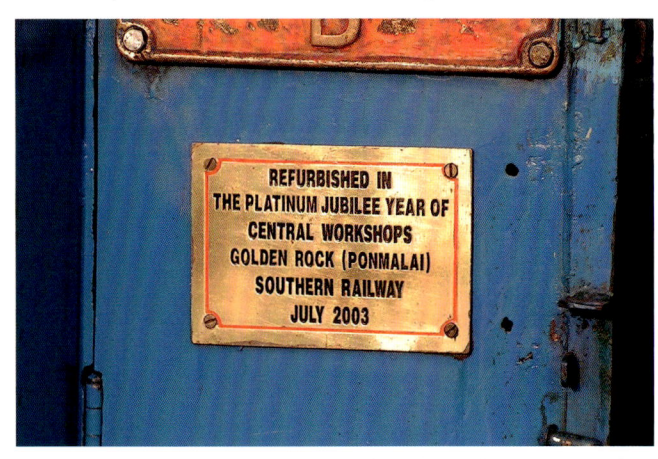

787 carried a Golden Rock rebuild plate during the oil firing trials.

Plates on 777 in 1971 showing the original style of Sharp Stewart plate and Tindharia rebuild plate. *(Michael Bishop)*

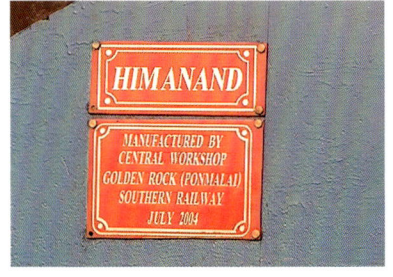

For a time, 780 had two names, one painted and the other on a brass nameplate. At Siliguri shed in 2006, the name WANDERER was being painted on, whilst it still carried the plate GREEN HILLS *(Alan Walker)*.

106

In 2002, No782 carried a Tindharia rebuild plate but no builder's plate. Note Hindi lettering on the usual right-hand side of the locomotive *(Andrew Young)*

797 in 1973 with Tindharia builder's plate. The lettering with overhaul data was frequently painted on the cab side *(Michael Bishop)*

Oil Firing the 'B' Class

Background and summary of systems used

The issue of global tenders in 1996 and 2000 for new oil-fired locomotives for the DHR is described on page 61 and Appendix 4. This section covers events after it was decided to produce the new locomotives in-house in India and give the work to the Southern Railway Central Workshops at Golden Rock, Tiruchchirappalli (abbreviated to GOC). TREC-STEP (an abbreviation for Tiruchirapalli Regional Engineering College Science and Technology Park) were also involved in the project, at least in the early stages. The work at Golden Rock continued for more than five years, during which four distinct oil-firing systems, identified as Stage 1, 2, 3 and 4, were trialled. Their salient features are summarised in the table overleaf. The forced draft system used in Stages 1 and 2 was highly unusual in steam locomotive applications.

STAGE 1: The First GOC/TREC-STEP system

'B' Class 787, taken from service on the DHR, was sent to Golden Rock in May 2002 for stripping down and installation of the first oil firing system for proving trials. The first system used high-speed diesel fuel supplied via two gear pumps to five LAP burners mounted at an angle of 40° to the bottom of the firebox. Forced draft combustion air was supplied to the burners by two centrifugal fans. The pumps and fans were directly driven by two Greaves model 4360 6.4 HP diesel engines. This equipment was mounted under the bunkers and not screened in any way. Air and fuel control was described as *'being by manual gate valves with manual ignition via the fire-hole door'*. Equipment for air braking was provided with large reservoirs slung on either side of the saddle tank.

787 was back at Siliguri for short-lived trials in November 2002 and was returned to Golden Rock that December. Press reports had 'sources in the railways' saying that the locomotive had to be sent back *'as the flames emanating from the diesel boiler leaped backwards when the locomotive headed uphill'*.

787 as it was as a 'Stage 2' loco at Siliguri Junction shed on 4 February 2004. The large air reservoirs for the brakes attracted some criticism and were subsequently concealed *(Peter Jordan)*.

Problems experienced during the trials were:
- The inadequate air supply only allowed use of three of the five burners – hence insufficient steam
- Poor atomisation due to low oil pressure producing black smoke. Manual control of the air/fuel ratio was laborious and difficult.
- Manual ignition through the fire-hole door was problematical, also 'back firing' and oil spillage, since the bottom of the firebox was open.

It did, however, complete a trial run from Siliguri to Km 10, which took six hours, eight minutes, of which four hours 42 minutes were spent waiting for the pressure to recover. The results of the trial are summarised on page 112.

STAGE 2: Modified GOC system

After it returned to Golden Rock in December 2002, 787 did not reappear at Siliguri for about 12 months. The oil-firing

	STAGE 1 First GOC/TRECSTEP	STAGE 2 Modified GOC	STAGE 3 BHEL	STAGE 4 Ffestiniog
Worked on by GOC	May 2002 to Jan 2003	Feb 2003 to Feb 2006	Sep 2004 to Apr 2005	Nov 2005 to Aug 2007
Loco No	787	787 1001 and 1002	1002	1001
Burners number and type	5 pressure jet	4 pressure jet plus pilot	4 steam atomised	1 steam atomised
Oil supply	Pump low pressure	Pump high pressure	Gravity feed	Gravity feed
Combustion air	Forced draft from fan	Forced draft from fan	Normal locomotive "induced" draft	Normal locomotive "induced" draft
Pump / fan power	Direct drive from small diesel engines	Electric from diesel gen-sets	None for burner, but on-board gen set for start-up air and brake	None
Other notes	Manual control	Electric control system		

On 25 November 2004, 787 is being worked on with modifications to extend the smokebox *(Peter Jordan)*

system was drastically changed with four ISA pressure-jet burners mounted at 90 degrees to the closed bottom of the firebox, plus a pilot burner with spark ignition. The ancilliary equipment was now electrically driven, including two centrifugal fans to supply forced draft combustion air, two gear fuel pumps, a feed water pump and compressors for the air brake system. The electric power came from two diesel powered gen-sets. The fuel supply to each burner was controlled by an on/off solenoid valve. The equipment under the bunkers was now screened for noise reduction and appearance. (A schematic diagram of the arrangement is shown by the drawing on page 109).

787, with this burner system, was trialled at Siliguri in January 2004. John Bancroft witnessed the tests and his observations were published in *The Darjeeling Mail*. Some extracts are given below.

24 January 2004: The generators were started and the fire lit. A clear exhaust was noted coming from the chimney, but also smoke leaks from around the smokebox. Steam was raised in about 30 minutes. The noise of the generators was deafening. Conversation was difficult within 10 metres of the loco and difficult even behind closed doors in the offices nearby.

30 January 2004: The actual test run. The loco was in steam when we arrived but, before coupling to the train, a section

Schematic diagram of Stage 2 oil firing system

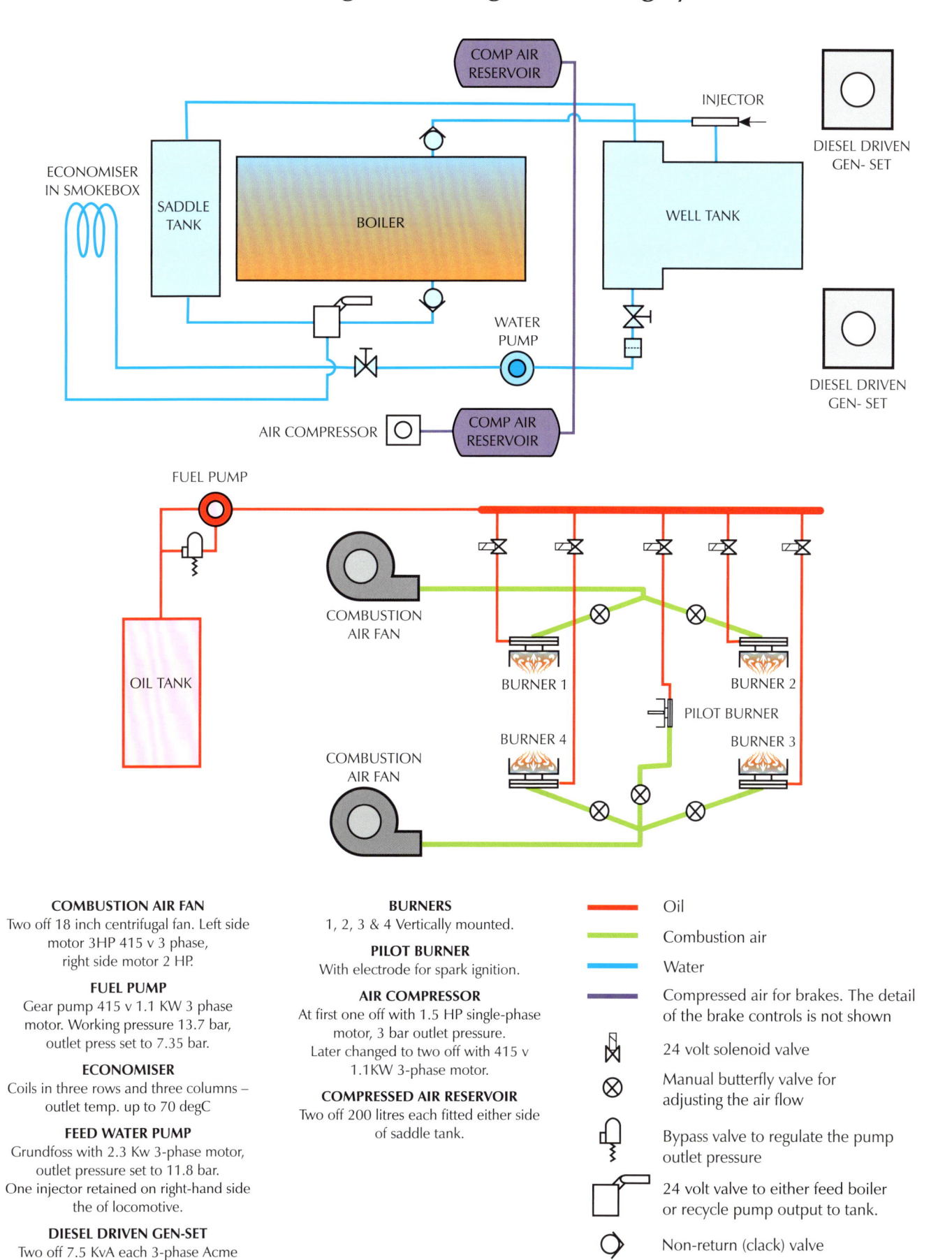

COMBUSTION AIR FAN
Two off 18 inch centrifugal fan. Left side motor 3HP 415 v 3 phase, right side motor 2 HP.

FUEL PUMP
Gear pump 415 v 1.1 KW 3 phase motor. Working pressure 13.7 bar, outlet press set to 7.35 bar.

ECONOMISER
Coils in three rows and three columns – outlet temp. up to 70 degC

FEED WATER PUMP
Grundfoss with 2.3 Kw 3-phase motor, outlet pressure set to 11.8 bar. One injector retained on right-hand side the of locomotive.

DIESEL DRIVEN GEN-SET
Two off 7.5 KvA each 3-phase Acme Motor engine fuel consumption 2.25 l/hr

BURNERS
1, 2, 3 & 4 Vertically mounted.

PILOT BURNER
With electrode for spark ignition.

AIR COMPRESSOR
At first one off with 1.5 HP single-phase motor, 3 bar outlet pressure. Later changed to two off with 415 v 1.1KW 3-phase motor.

COMPRESSED AIR RESERVOIR
Two off 200 litres each fitted either side of saddle tank.

— Oil
— Combustion air
— Water
— Compressed air for brakes. The detail of the brake controls is not shown

24 volt solenoid valve

Manual butterfly valve for adjusting the air flow

Bypass valve to regulate the pump outlet pressure

24 volt valve to either feed boiler or recycle pump output to tank.

Non-return (clack) valve

© David Churchill 2018

109

of the extended cab roof had to be cut off to prevent it fouling the coaches. The train left Siliguri Junction at 11.38, only to stop for about 10 minutes outside the shed. It then proceeded to Sukna. The exhaust was clear except for some black smoke when the engine shut off. General speed was about 15 km/h but the exhaust was very loud and laboured for a locomotive pulling only 1.2 times its own weight on near-level, straight track. Arrival at Sukna was at 12.30. The loco blew off within a minute of arrival, over a full glass of water. We departed Sukna at 12.57; the exhaust was very loud and laboured, with an estimated maximum speed of 10 km/h on the gradient – at times it dropped below walking pace. We stopped just over one road crossing – engine had half a glass of water in the boiler and full pressure. After more adjustments had been made, we restarted, progress being at walking pace. We finally stopped at Km 15.12 at 13.52. The locomotive had full pressure, but little water showing in the gauge glass even on the steep up-grade and an empty water tank. The train then backed down to Sukna, and was eventually pulled back to Siliguri Junction by another locomotive.

787 was further modified and tested on several occasions between 2004 and 2006. The modifications included changing to a large saddle tank with the compressed air reservoirs for the air brake concealed rather than being slung externally on either side. Trials in February 2006 were observed by David Barrie and David Wilding. David Wilding's observations were published in *Darjeeling Mail* and some extracts are included below.

White cars followed the train with the guests and dignitaries. We drove ahead to catch the oil-fired locomotive and caught up with it just after the fourth road crossing in the forest. It was stopped with a big crowd around it – feed water problems. After lots of valve turning, pump bleeding and feeling the temperature of the feed pump motor, the injector on the other side was cranked back into life and after a short wait for boiler fill and steam raising they were off again.

787 arrived looking very strong at the old loop No1 , now replaced by a 1 in 14 gradient – I know it is 1 in 14 because I measured it with a calibrated spirit level I made for this visit! 787 slowed on the steep bit, but slowly ran out of steam until it stopped on the 1 in 14 section. After a few minutes to raise steam it pulled away without much fuss, just the one wheel turn of slip. There was a noticeable lack of safety valve lifting or the leaking steam we had got used to. It was a little boring! The Safari Special passed at the same spot, not long after, and only just made the 1 in 14 without stalling. Both locomotives had three coaches behind them.

Cab view of 787 in February 2006 after the air reservoirs had been concealled. It is very different from the normal 'B' Class cab, with its wide-screen front window, row of electrical indicator lights under the cab roof and no bunker for the coal feed. Note some parts associated with the air brake originally installed: its extended brake shaft, reservoir and some pipework.

We drove forward to Rangtong to watch the arrival there. 787 arrived in the passing loop. A crowd gathered around the left-hand side of the loco, the hammer came out and some fine adjustments were made. Then people started drifting over to the white cars and drove away.

We filmed 787 pull away then drove up to the water tower. 787 was a long time coming. When it did arrive the hammer came out again and some serious adjusting took place. The problem was that the left side piston rod had come loose in the crosshead so the hammering was to knock the crosshead retaining wedge into the piston rod.

After filling the loco with water, the carriage brakes were adjusted for a slow run downhill back to Siliguri Junction.

The day after, we saw 787 and its train of three coaches at Tindharia having steamed up the hill without a problem. We followed it down the hill later and caught up with it at Sukna. I had a close look at it and saw the smokebox feed water element had been disconnected and all the associated valves were in the off position, so they must have been running on the injector again.

The results of the final 2006 trial are summarised in the table on page 112.

After 2006, 787 stood out of use at Siliguri for some time before eventually being moved to Tindharia and gradually stripped of equipment. Meanwhile, two new locos, and possibly a third, had been built at Golden Rock, at first to an apparently similar 'Stage 2' design.

1001 HIMRATHI was built and flagged off in July 2003. It remained at Golden Rock until January 2007, by which time it had been fitted with the 'Stage 4' Ffestiniog burner system.

1002 HIMANAND was completed in 2004 at a cost of Rs.1.5 crores, with certain modifications. However it did not get to Siliguri until March 2005, after it had been fitted with the 'Stage 3' BHEL burner system.

It is not clear whether the third locomotive, No 1003, was ever completed, although a press report does suggest that it was. If it in fact was built, it seems never to have left Golden Rock.

Pictured on 21 February 2006, 787 has successfully made a run to Tindharia *(Alan Walker)*.

Cedric Lodge in action at Siliguri Shed on 16 February 2007 *(Dave Priestley)*.

Summary of Trial Results

Burner system	Stage 1	Stage 2	Stage 3	Stage 4
Date of trial	October 2002	February 2006	April 2005	7 Auguust 2007
Loco no	787	787	1002	1001
Load	2 coaches	3 coaches	……………	3 coaches
From / to	NJP to Km 10/0	SGUJ to TDH	Within SGUJ shed	SGUJ to Km 19/14
Total time taken	6 h 8min	4 h 13min	…………	2h 48 min
Stops for press build up	10	8	………………	6
Idle time for stops	4h 42 mins	47 min	………………	38 mins
Press fall to	3.5 kg/cm2	6 kg/cm2	………………	7 Kg/cm2
Oil consumed	240 litres	12 litres /km	………………	640 litres
			Steam generation not adequate for traction	

STAGE 3: BHEL gravity fed system

In April 2004, after feedback from trials of 787 at Siliguri, Bharat Heavy Electricals Ltd were approached to design and manufacture a gravity-feed oil system. They developed a novel oil burner described as 'gravity-fed' with *'combustion air generated by steam ejectors in the chimney'*. They applied for a patent for the burner in July 2004, which was granted in 2010. The newly-built No 1002, which was still at Golden Rock, was fitted with four of the BHEL steam atomised burners. Air brake reservoirs and compressors were now hidden and new economisers introduced in the smokebox to enhance the efficiency of the boiler. The saddle tank water capacity had been increased from 1,800 to 2,100 litres, with a 900-litre capacity diesel tank. A 5KW generator provided power for lighting and the air brake compressor. The sound level was reduced and it was expected that the

1002 Himanath in February 2006, looking unloved, a short while before its return to Golden Rock

loco would haul two to three coaches at a speed of 20km/h in the hilly terrain.

1002 arrived at Siliguri in March 2005 and had a short trial at Siliguri shed in April. It was recorded that steam generation was not adequate for traction as only two burners could be used due to inadequate air supply. The steam atomisation was wasteful and the burners having, by design, to be mounted vertically, caused the flames to impinge on the crown plate. It seems that nothing further was done to 1002 and it finally returned to Golden Rock in March 2006. Press reports at the time said that it was *'returned due to nozzle problems – not yet commissioned into service'*. Nothing further has been recorded of 1002 since (see photo on page 93).

STAGE 4: Ffestiniog gravity fed system

The Executive Director-Heritage of the Railway Board suggested to Golden Rock that the proven oil firing system used by the Ffestiniog Railway should be fitted to a 'B' Class.

1001 which had remained at Golden Rock since it was built in 2003 was the locomotive chosen. It eventually reached Siliguri in Jan 2007, fitted with the burner system used by the Ffestiniog Railway who had provided the outline design and some vital components. Light burning oil was fed by gravity to a single burner in the firebox, where it was atomized by the use of steam from the boiler (or air from an outside supply whilst raising steam) There was no need for additional burners, fans or gen-sets. A 220 gallon fuel tank was designed to fit into the B class bunker with a balancing pipe between the two lower sides of the tank to enable the volume available to be fully used. A drain valve was included to enable water contaminant to be drained from the bottom of the balancing pipe.

There were 3 controls: oil, atomising steam and blower. For successful oil firing, it was advised that it was essential that the fireman be able to see the chimney in order to trim the flame to best effect. Trimming was to be achieved by adjustment of the 3 controls to produce an exhaust described as 'a light haze'

It was trialled under the supervision of Cedric Lodge, from the Ffestiniog Railway in 2007. Some of his observations on the tests are included as Appendix 5. Cedric reported to Indian Railways in detail on his work and it was understood that the intention was that he would return for further trials after some of the shortcomings had been addressed. However this never materialised. A further test was completed in August 2007 and witnessed by a number of

1001, by now without its name, looks immaculate externally at Siliguri Junction shed *(Dave Priestley)*.

A cab view of 1001 fitted with the 'Ffestiniog' oil-firing system. There are additional controls for oil firing on the left-hand side where the outlet from the coal bunker would normally be. The Detroit-type lubricator has had to be mounted higher up than usual *(Dave Priestley)*.

After the oil firing trials, 787 remained at Siliguri Junction for a while before being taken to Tindharia works where it was gradually stripped of useful parts and left to decay. It is shown here on 31 May 2015. It is proposed to resuscitate it as a coal-fired locomotive and work appeared to have begun by early 2018. *(Nick Fitzgerald)*

important railway officials. It is not clear to what extent the issues raised by Mr. Lodge had been attended to but similar problems were reported after this trial including lack of steam because full burner capacity could not be used due to inadequate air supply and difficulty with injectors. The results are summarised below, although the figure quoted for oil consumption does seem to be extremely high.

Probably influenced by the trial it appears that a decision was taken in autumn 2007 to terminate the oil firing work on the DHR. At the time it was also reported that the coal supply situation had improved. 1001 was converted to coal firing by November 2007.

However this was not quite the end for the oil fired B class!!

Oil Firing at Matheran

'B' Class 794 had been at Matheran for some time, although very seldom used. In 2013 Central Railway decided to convert it to oil-firing because of environmental concerns, in particular lineside fires in the eco-sensitive Matheran region. It went to Golden Rock for conversion in April 2013.

The work took 30 days and it was despatched at the beginning of June, having been 'flagged off' by the Director, National Institute of Technology. It returned to Neral, following which the CR began trials. In April 2014 it was reported as 'not in a position to run' but it was hoped it could be operational by that October. An official was quoted as saying 'replicating the model (ie Nilgiri) in the Neral -Matheran section has proved 'adventurous'.

The conversion provided the loco 'with twin-head burner, combustion blower with air control, fuel pump, electrical control unit, fuel oil system accessories, and diesel oil tank'. The oil tank (capacity 800 litres) was installed in the bunker area. A 5 KVA electrical generator set was installed on the side platform of the loco. The water capacity was quoted as 1,818 litres. This appears to be yet another different system – a variation on the 'Stage 2' GOC system, with pressure oil feed, forced draught air supply and air fans and pumps electrically powered from diesel-driven gen-sets. It was given a fresh coat of polyurethane paint. Anticipated performance was given as 'will run at a speed of 13km/h hauling four coaches'.

The locomotive appears to have been hardly (probably never) run at Matheran until December 2017, when it was reported as having run trials with the aim of running some steam services on the line, newly re-opened after being closed for some time due to safety concerns. In April 2018 it made a well-publicised inaugural run from Matheran to Aman Lodge with two coaches, which was reported in the press as being successful.

These two pictures, taken 54 years apart, highlight just how DHR has changed and no doubt this will continue. One constant feature has been the 'B' Class and, whatever the future might hold, they must remain a working part of it

[Top] In their prime the 'B' Class could haul four coaches and a van up to Darjeeling: now they are limited to just two coaches. *(The late Peter Bawcutt courtesy of Peter Tiller)* and [Above] 'Freight' charters for photographers were a new innovation in 2017, well illustrating the fact that the railway is now very much in the tourist rather than the transport business! Here 802 pauses for water above Kurseong *(Peter Jordan)*

Appendix 1

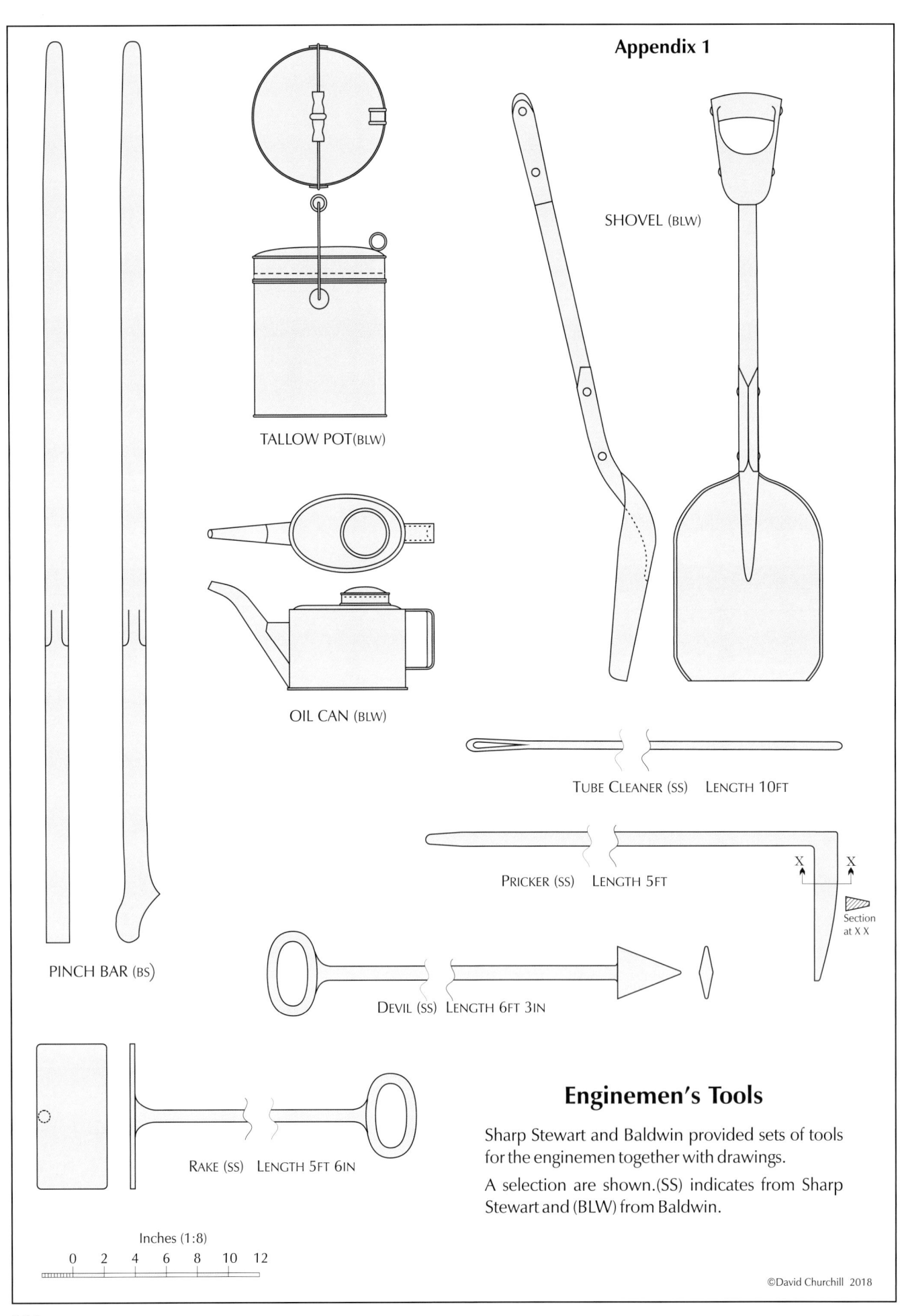

SHOVEL (BLW)

TALLOW POT (BLW)

OIL CAN (BLW)

TUBE CLEANER (SS)   LENGTH 10FT

PRICKER (SS)   LENGTH 5FT

Section at X X

DEVIL (SS)   LENGTH 6FT 3IN

PINCH BAR (BS)

RAKE (SS)   LENGTH 5FT 6IN

Enginemen's Tools

Sharp Stewart and Baldwin provided sets of tools for the enginemen together with drawings.

A selection are shown. (SS) indicates from Sharp Stewart and (BLW) from Baldwin.

Inches (1:8)
0 2 4 6 8 10 12

©David Churchill 2018

Appendix 2   Sharp Stewart Specification for 'A' Class

Reproduced from a transcript *'ex-Halfway to Heaven'* by Terry Martin courtesy Fabien Raymondaud, Terry Martin Estate, and John Milner, Rail Romances Publishing. Original material courtesy Don Townsley.

Four of the 'A' Class locomotives were built by Hunslet under sub-contract to Sharp Stewart, and Sharp Stewart provided a specification to which they were to be manufactured. It provides a useful insight into materials and practices in the 1880s. It is directly relevant to the 'B' Class as Sharp Stewarts order book for the first 'B' Class said in details generally follow E810, which was the second order for 'A' Class.

Atlas Works Manchester Specification of Locomotive Tank Engine. To drawing No. 8995 Gauge of Railway 2 feet 0 inch

Cylinders. The cylinders are to be outside the frames and cast of strong close grained metal as hard as can be bored and planed. All the joints are to be fitted metal to metal. The slide valves and glands are to be gun metal and each cylinder is to be provided with two lubricator and three waste water cocks.

Pistons. The pistons are to be made of cast iron and the packing rings of cast iron.

Wheels & Axles. The wheels are to be made of solid wrought iron throughout, and the tyres of Bessemer steel, wheel centres are to be discs blocked into form under the steam hammer. The axles are to be made of Bessemer steel and to have outside crank arms of wrought iron. The crank pins, which are to be well fitted into the crank arms, are to be secured by riveting and to be of Bessemer steel. The wheels are to be counterweighted where required.

Axle Boxes. The axle boxes are to be of cast iron and the steps of gun metal well fitted in.

Frames. The frames are to be outside the wheels and made of wrought iron plate equal to best Staffordshire quality. They are to be well connected transversely by suitable stretchers of wrought iron plate and angle iron. They are to be set at the hind and outwards to allow of an extra width of fire box. The rivets connecting the fire and hind joints of the frame to be turned and the holes rhymered and the riveting to be very carefully done.

Axle Box Guides. The axle box guides are to be of strong cast iron fixed by turned bolts fitting tightly into rhymered holes.

Springs. The springs are to be of the best spring steel and the links and brackets of wrought iron.

Buffers & Draw Gear. The buffers at each end to consist of buffing plates of wrought iron on India rubber pads curved to suit the rounded frame work of the ends of the engine. The draw gear of each end is arranged with a radial bar and links.

Splashers & Sand Boxes. Neat splashers are to be fixed over the wheels where requisite, also sand boxes to be supplied and gear for working them from the foot-plate.

Safety Chains. Safety chains are to be provided at each side of the draw bars for both ends of the engine.

Awning. An awning is to be fixed over the foot-plate which is to consist of sheet iron suitably stayed and carried by turned wrought iron pillars. Two lookout glasses in brass frames are to be provided in the front plate.

Tool Box. A tool box of wrought iron plate is to be fixed at the front end of the engine.

Boiler & Fire Box Shell. The boiler is to have the longitudinal seams lap jointed and double riveted, each ring in one plate. The circular seams are to be lap jointed and single riveted. The front tube plate is to be flanged to meet the smoke box plates. The fire box shell front and back plates are to be flanged to meet the sides and top, and the shell and boiler are to be connected by a flange on the front plate of the shell. On the boiler is to be riveted a dome having a joint at the upper part, and on the top of the fire box shell, a wrought iron man hole ring is to be riveted. All the plates and angle irons of the boiler, dome and fire box shell are to be of Bowling iron. At each side of the fire box shell is to be riveted a long angle iron through which the fire box shall rest on the frames, and to be free to slide as the boiler expands longitudinally, the other end of the boiler being fixed through the smoke box to the frames. The back plate of the fire box shell is to be well stayed to the boiler body and front tube plate by longitudinal stays. The rivets are to be of Low Moor or Bowling iron – ¾ of an inch diameter and 1¾ inches apart centre to centre.

Tubes & Fire Box. The tubes are to be of solid drawn brass; they are to be carefully fixed in the tube plates, and are to have steel ferrules driven tightly in the fire-box end only. The fire-box is to be made of the best copper; the front, sides and back are to be stayed to the fire box shell by copper stays 7/8 in. diameter, and 11 threads per inch, screwed tightly in and riveted over at the ends, the distance apart centre to centre to be about 4 inches, the bottom of the fire-box is to be attached to the shell by a solid wrought iron ring, and rivets 7/8 in. diameter, 2 ins. apart centre to centre. The roof is to be stayed by wrought iron stays screwed into the copper plate of the fire-box and the iron plates of the shell being riveted over outside the latter and having nuts above and below the copper plate; also three transverse screwed stays above copper box. The roof of the copper box is to be inclined from tube plate to the back downwards so that the back part may not be bare of water when the engine is descending the inclines. The fire bars are to be of wrought iron and supported by strong bearers secured to the bottom ring of the box.

Safety Valves. Two safety valves are to be fixed on the fire-box shell. They are to be of "Ramsbottom" system and adjusted to a working pressure of 140lbs per sq.inch and to be fitted with wrought iron funnels to conduct steam above awning.

Regulator. The steam regulator is to be placed in the dome and to have a pipe leading to the smoke-box, and on the fire-box back a brass stuffing box for the rod, and stops for the lever.

Fire Box Fittings. A "Bourdon's" pressure gauge with a tap on the fire-box, a glass water gauge, two gauge cocks, whistle and a blow-off cock are to be provided; also a fire-box door and ash-pan with dampers worked from the footplate; brass wash-out plugs are to be screwed in the bottom of the fire-box shell at the corners where practicable, also in the back of the shell level with the crown of the fire-box, and in the smoke-box tube plate above and below the tubes.

Smoke-box & Chimney. The smoke-box and chimney are to be made of wrought iron plate equal to best Staffordshire quality. The chimney top is to be cast iron. The smoke-box is to have a circular dished door and liner plate, also the requisite handles, fastening and hinges. The steam pipe leading to the cylinders is to be of copper, and the exhaust pipes of wrought iron with a copper top. A steam jet cock worked from the foot-plate, with pipe leading into the chimney is to be provided.

Motion Work. The connecting and coupling rods are to be made of the best hammered scrap iron filled with gun metal steps. The slide bars are to be made of the best hammered scrap iron well case-hardened. There is to be one only to each cylinder. The piston road crossheads and pins are to be made of the best hammered scrap iron, with slides of gun metal and the piston rods of Bessemer steel. The whole of the link motion is to be of the best selected scrap iron well case-hardened throughout. "Walschaerts System" to be outside the frames. The eccentrics are to be of cast iron. The reversing is to be by lever with sections which are to be fixed on the right hand side of the engine.

Injectors. The boiler is to be fed by one brass injector No.5 size, of our most approved make and having a back pressure valve at the foot and one on the boiler of brass.

Pump. A pump is also to be provided; it is to be made of brass and worked from the crosshead. All the pipes for both pump and injector are to be of copper. Suitable clacks and taps are to be provided.

Tank & Fuel Box. The tank is to be fixed underneath the boiler and smoke-box and to be made of plates equal in quality to the best Staffordshire iron, and riveted together by '/> inch rivets cup-headed on the outside; the sides are to be well stayed, and a suitable manhole with lid and sieve are to be provided. The fuel boxes are to be placed one on each side of the footplate and to be made of the same quality of plate as the tank and to be supplied with a door sliding in grooved guides. Handrails are to be provided where required.

Break. The break is to be worked by a screw and handle from the left-hand side of the foot-plate and is to act on one side of each wheel. The blocks are to be of cast iron.

Skid Break. A skid break is to be provided, the C.I. blocks arranged for bearing on the rails when in action at each side of the engine, are to have cast iron shoes removable for renewal.

Finishing Painting & Tools. The boiler and fire-box to be clothed with wood and sheet iron, and the corners of the firebox to be finished with charcoal iron moulding. The dome casing is to be of charcoal iron. The engine is to be finished in the neatest possible manner and in every respect complete and ready for work. The workmanship is to be of first class character throughout, and the materials used are to be of the best description of their respective kinds, perfectly free from any defects whatever. The screws are to be "Whitworth's" standard. All the parts usually painted are to be finished, with such colors as may be chosen. The engine is to be tried in steam to a pressure of 140lbs per square inch, and the boiler to be tested by hydraulic pressure to 200 lbs. per square inch. The engine is to be provided with one set of fire irons, one shovel, one set of screw keys, one hand hammer and chisel, one oil bottle and feeder, one Tommy bar and one monkey wrench. After the engine has been tested, it is to be taken in pieces and the parts packed in strong packing cases, all bright work being carefully coated with white lead and tallow, or other preparation, and the journals of axles and similar parts securely lapped round. All parts to be perfectly clean and free from rust before any coating is applied. The whole of the work that may require it is to be protected by painting or otherwise.

The following parts will be supplied by Sharp Stewart & Co Ltd. -gratis. Wheel centre forgings, chimney bases, all iron and brass castings, injectors finished.

Appendix 3
Manufactured paints for modellers

For guide only. All colours are subject to the limitations of colour printing.

Phoenix Paints P90 SR Venetian Red	Phoenix Paints P12 BR Freight Wagon Bauxite
Vallejo 71/06 Brown	Rail Match GW Indian Red
Tindharia Pot 1	Tindharia Pot 2
Phoenix Paints RAL5018 Capri Blue	Phoenix Paints RAL3009 Azure Blue
Vallejo 70962 Flat Blue	Vallejo 70841 Andrea Blue
Vallejo 71005 Intermediate Blue	Vallejo 70840 Light Turquoise
Phoenix Paints P750 early GNR Loco Green	Phoenix Paints P751 late GNR Loco Green

Appendix 4   Notes on DLM proposal

Roger Waller of DLM has kindly allowed inclusion of the following paper which describes some background and technical details of DLM's proposal against the Global Tender of 2000 for new locomotives for the DHR.

DLM-Project for New Steam Locomotives for Darjeeling

ROGER WALLER, Dipl.-Eng. ETH

1. Introduction

In 1952 the Swiss Locomotive- & Machine Works SLM delivered a further batch of new 0-8-2 T Rack- and Adhesion Steam Locomotives to the Indian Nilgiri Mountain Railway. These five X-class locomotives (X13 – X17) were not much different from the previous series built in 1914 (X1 – X4), 1920 (X7 – X10) and 1925 (X11, X12).

SLM declared these five X-class locomotives to be the last steam locomotives built in Switzerland, but history proved them wrong. 40 years later, on the initiative and under the direction of the present author, SLM built three prototypes of a modern, all new rack steam locomotive. One-man operation, all-welded light-weight construction, ultra-clean oil-firing, roller bearings and electronic safety devices greatly improved the economics and reduced the ecological footprint to a level that was unknown for traditional steam locomotives. Following the success of the prototypes, a series of five more modern rack tanks were built in 1996. SLM chose the branding of **modern steam** to get rid of the poor image of the old steam locomotives as being out-dated, old-fashioned, smoky and dirty. For the philosophy, history and development of **modern steam** see [1].

Switzerland with its fully electrified railways was certainly the most unlikely country to restart steam locomotive development. The news of modern steam locomotives being built in the renowned Swiss Locomotive- & Machine Works in parallel to the latest high power electric locomotives with radial-steered axles received worldwide attention. Had the same work been done in a third-world country, it would have hardly been noted.

2. Nilgiri Mountain Railway

This metre-gauge rack and adhesion railway was operated exclusively by SLM steam locomotives. Due to their old age and a rather limited maintenance it was necessary to renew the fleet. In September 1992 the Southern Railway, a division of Indian Railways, contacted SLM to find "an alternative mode of traction on this mountain railway system". This enquiry initiated various engineering studies for diesel railcars and locomotives. Even electrification had been discussed! Not surprisingly electrification proved to be too expensive. Diesel railcars did not provide sufficient transport capacity.

Much engineering and design work was then done by SLM for new diesel locomotives, but it was soon found that the conditions on this line (restricted clearance profile, axle load of only 10.7 tons, weak Abt rack bars) did not favour diesel traction. But the real crux was to find a suitable solution for a combined rack- and adhesion drive on diesel locomotives. SLM had recently developed a so called *Differential Drive* for electric locomotives type HGe 4/4 II sold to several Swiss metre-gauge rack- and adhesion railways. However, with an axle load of 16 tons these were far too heavy for the Nilgiri line. Further the differential drives were a very sophisticated, complicated type of gear and were not considered an appropriate technology for the rather basic maintenance workshop at Coonoor. So even the toughest promoters of diesel traction had to admit that steam traction in this case was far superior, even on pure technical grounds.

Following a presentation of SLM's **modern steam** technology to the Indian Railway board on 30 March 1995 at New Delhi, Indian Railways decided to shelve the diesel and electric options and concentrate on new steam locomotives. The strategy of Indian Railways was now to purchase four new oil-fired steam locomotives for the Nilgiri line and three new oil-fired steam locomotives for the Darjeeling line. As a basis for global tenders, technical specifications were worked out for both projects.

3. Global Tenders for new steam locomotives

In 1996 a global tender had been issued by Indian Railways:

GP 186: four new oil-fired rack- and adhesion steam locomotives for the Nilgiri line.

This was soon followed by a similar tender for three new oil-fired steam locomotives for the Darjeeling Himalayan Section.

SLM decided to initially concentrate on Nilgiri. SLM were the only company to put in an offer for these sophisticated locomotives. No offers were received for the Darjeeling steam locomotive tender!

Indian Railways were happy with the technical bid of the SLM proposal, but less so with the commercial bid. The quoted price was higher than the budget of Indian Railways for these locomotives working on a branch line that in a general Indian context was not considered as important. Three factors were relevant for the seemingly high price:

1. Combined rack- and adhesion locomotives are technically most complex with four cylinders, reduction gears and three independent brake systems.

2. The small number of locomotives prevents economies of scale. If series of several hundred locomotives can be built, engineering and design costs per locomotive become almost negligible. Not so, if you have to divide them by four.

3. India at that time had very high import taxes (some 74%) on locomotives. Adding the costs for transport, insurance and financing meant that Indian Railways had to pay around double the price of what SLM actually asked for. These costs were beyond the influence of SLM. Of course the Indian representatives of SLM tried to waive the import taxes, but did not succeed.

Even so there was no countable result from these Global Tenders, correspondence and talks between Indian Railways and SLM continued.

4. The end of the Swiss Locomotive & Machine Works SLM

In June 1998 Sulzer Ltd., the main shareholder of SLM, sold the engineering of SLM to Adtranz. The workshop of SLM was renamed Sulzer Winpro. Adtranz had no interest in the steam engineering so that it became part of Sulzer Winpro. It was clear that there was not much future for Sulzer Winpro as a simple workshop without engineering and those were not interested in continuing the steam business. However, the two important contracts for a new 650 kW steam engine for the paddle steamer "Montreux" and the modernization of the steam locomotive 52 8055 secured our steam activities for another two years. Then it was time for a management buy-out which led to the foundation of the Dampflokomotiv- und Maschinenfabrik DLM Ltd. (www.dlm-ag.ch).

5. More Global Tenders for new steam locomotives

In the year 2000 Indian Railways basically re-issued the two global tenders from 1996:

GP 192: three new oil-fired steam locomotives for the Darjeeling Himalayan Section.

GP 193: four new oil-fired rack- and adhesion steam locomotives for the Nilgiri line.

Being a smaller company than SLM, DLM decided to offer the Darjeeling locomotives only. Indian Railways were informed that

119

DLM were interested in the Nilgiri contract as well, but would not have the capacity to handle both projects at the same time. The result was that no one responded to the Nilgiri tender. This time three companies participated in the Darjeeling tender:

1. Dampflokomotiv- und Maschinenfabrik DLM Ltd., Switzerland
2. Rheilffordd Ffestiniog Railway, UK
3. Alan Keef Ltd., UK

The bid documents had to be sent in two separate folders, one for the technical bid and one for the commercial bid. First the technical bids were opened and selected. The bid of the Ffestiniog Railway was incomplete and not accompanied by the compulsory bid guarantee which led to the immediate disqualification of their bid. Alan Keef's bid was considered, but not selected on technical grounds, mainly for lack of references. This left DLM as the only company to have passed the technical selection. The following are extracts of the technical specifications that were part of DLM's technical bid:

UNIQUE INNOVATIVE FEATURES OF STEAM LOCMOMOTIVES MANUFACTURED BY DLM/SLM

SLM/DLM has developed a new steam locomotive technology which resolved the economical and ecological problems of traditional steam locomotives. This new technology which nowadays is *exclusively* in the hand of DLM Ltd. is named **modern steam**.

Main advantages of the **modern steam** – technology are:

- **Oil Firing** with light oil or high speed diesel fuel
- **Smokeless Combustion** with excellent emission values outperforming the ones of diesel engines
- **High Efficiency** of the steam locomotives thanks to improved cycle efficiency (higher steam pressure, internal streamlining, higher degree of superheating)
- **More Power** thanks to better design of boiler and steam engine
- **Higher Speed** thanks to better balancing and more power (especially important to increase the uphill speed)
- **Short Preparation Times** thanks to various improvements like oil firing, efficient insulation, roller bearings and careful detail design
- **Reduced Staff Requirements**, equal to the requirements of Diesel or electric locomotives
- **Minimal Maintenance** thanks to modern technology
- **Low Fuel Costs** thanks to high efficiency of boiler and steam engine
- **Simple Operation and Maintenance** thanks to rugged construction

The application of **modern steam** – technology to newly designed steam locomotives means highly competitive locomotives in the economic as well as the ecologic sense. The locomotives do however retain the traditional appearance and the aesthetics of the steam locomotives, so much liked by everybody.

TECHNICAL DESCRIPTION OF THE NEW DARJEELING STEAM LOCMOMOTIVES

1. INTRODUCTION

The new steam locomotives for the Darjeeling line of the Indian Railways are a new design of the Swiss Dampflokomotiv- und Maschinenfabrik DLM Ltd. They follow the principles of **modern steam**, but are similar in appearance to the existing "B"- class locomotives. Technically they are based on the modern rack steam locomotives supplied in 1992 and in 1996 ….

2. OPERATING CONCEPT

The new concept of these modern steam locomotives eliminates the well known disadvantages of traditional steam traction.

2.1 Engine Crew

The locomotive crew consists of a driver and a fireman only.

2.2 Turnaround Time

The turnaround time is very short. There is no need for fire cleaning and neither the ashpan nor the smokebox have to be emptied. Only water has to be taken, otherwise the turnaround time is equal to the one of diesel or electric traction.

2.3 Preparation Time

2.3.1 Electric Preheating Device

The locomotive can be equipped with two connections to fit an external electrical preheating device which heats the cold boiler water unattended. By virtue of the oil firing, the locomotive steamed up in this way is ready in a few minutes. The slow and uniform heating reduces the stress on the boiler.

2.3.2 Standby Losses

Old steam locomotives have considerable radiation losses for lack of insulation. These results in high energy consumption and a significant drop of boiler pressure while on standby (e.g. overnight). The new steam locomotives have efficient boiler insulation which reduces the radiation losses to a large extend. With a chimney cover and closed air dampers, pressure drop is less than 8 bar in twelve hours. If the locomotive is used daily, the electric preheating device is not required. The boiler has sufficient steam to start the oil-firing directly.

Since radiation losses occur consistently when the locomotive is in steam, a good insulation saves a lot of energy (and money!) throughout the year.

3. TECHNICAL DATA

3.1 Dimensions

Gauge:	610 mm
Length overall:	6540 mm
Maximum width:	2135 mm
Maximum height:	2895 mm
Driving wheel diameter (new):	830 mm
(worn):	90 mm
Wheel base:	1670 mm
Boiler pressure:	16 bar
Cylinder diameter:	280 mm
Stroke:	400 mm
Water capacity approx.:	2000 litres
Fuel capacity approx.:	700 litres

DLM Ltd. reserves the right to modify the dimensions during the design process if required.

3.2 Weights

Weight empty, approx.:	14'000 kg
Weight in running order (with 2/3 supplies):	15'500 kg
Maximum axle load (with 2/3 supplies):	7'500 kg +/- 2%

3.3 Fuel and Water Consumption

Fuel and water consumption will be as follows:

	fuel:	water:
New Jalpaiguri to Darjeeling with four coaches (30 t):	405 l	3780 l
Darjeeling to New Jalpaiguri with four coaches (30 t):	59 l	560 l
New Jalpaiguri to Darjeeling with five coaches (37.5 t):	468 l	4380 l
Darjeeling to New Jalpaiguri with five coaches (37.5 t):	66 l	620 l
Stand-by per hour:	1 l	10 l

3.4 Speed

On clean rails and no restrictions due to cant deficiency or poor track quality, the following speeds can be maintained:

Train weight (without locomotive):   30 t   37.5 t

- Level and uphill
  up to 20 ‰ equivalent gradient: 40 km/h   40 km/h
- Up to 30 ‰ equivalent gradient: 40 km/h   36 km/h
- Up to 40 ‰ equivalent gradient: 39 km/h   29 km/h
- Up to 50 ‰ equivalent gradient: 28 km/h   24 km/h

As can be seen, the new steam locomotives have plenty of reserve power which is important for the future, allowing higher speeds for parts with improved track.

Following the rules for Global Tenders set by Indian Railways, only the commercial bid of DLM was opened. It seems that the commercial bid was less convincing than the technical bid, because there was no order to follow. We considered our prices to be fair, even moderate in view of the considerable development, design and manufacturing work necessary to realise such an ambitious project. Indian Railways did not share this view which is no surprise considering they would have to pay twice of what DLM gets. As has been explained in the Nilgiri chapter, the main problem was the exorbitantly high Indian import tax. As for the Nilgiri project our Indian representatives had put in a request to waive the import tax but this was again not granted. DLM then tried to get support from UNESCO. We were informed that UNESCO already had far too many projects to cope with in view of their very limited resources. We also learned that the heritage status was given to the Darjeeling line, not to the steam locomotives and the other rolling stock. Chances of getting financial support from UNESCO for this project were declared nil.

DLM then got in contact to the Darjeeling Himalayan Railway Heritage Foundation, hoping they would support the case. After all, new steam locomotives would have secured the ordinary time-tabled rail service on the entire line with the bonus of replacing diesel locomotives. An attractive aim for a supporting railway society one would think. But again, there was no support at all. DLM then decided to have drawn enough blanks and that it was time to concentrate on other projects.

Indian Railways then made attempts for reverse engineering for both the Darjeeling and Nilgiri steam locomotives. However, the resulting replica steam locomotives are not very convincing, neither technically, aesthetically nor acoustically.

For the present author real steam means regular time-tabled service. New, modern DLM steam locomotives could have successfully outperformed today's diesel locomotives, as has been proven with the new rack tanks. With hindsight one can certainly say that an opportunity for sustainable real steam traction has been missed.

LITERATURE

[1] Waller, Roger: MODERN STEAM – AN ECONOMIC AND ENVIRONMENTAL ALTERNATIVE TO DIESEL TRACTION; Institution of Mechanical Engineers IMechE, Railway Division; Lecture presented at the Sir Seymour Biscoe Tritton Lecture on Monday 3 February 2003 and Tuesday 4 February 2003.

Appendix 5   Oil firing 1001. Cedric C Lodge CEng; MIMech E

During a gala weekend on the Ffestiniog Railway (FR) in April 2005, during which Minfordd was dressed up as Sukna (Darjeeling Mail, Issue 30) , it came to my attention that John Prideaux was donating a burner and Stordy oil control valve to Indian Railways, to enable them to equip one of the 'B' Class locomotives with the FR oil firing system. What was now required was somebody to undertake the detailed design of the oil fired system for application to the 'B' Class locomotive which Indian Railways would make and install. The challenge was irresistible, and I volunteered. I was delighted when my offer was accepted.

I had a General Arrangement (GA) of a 'B' Class locomotive. upon which to base the oil firing design. Living close to the FR enabled me to study details of the FR system and incorporate them into the design. The drawings, specification and operating instructions were sent to Indian Railways via DHL. Then everything went quiet for nine months, until I received a request from Indian Railways for a design of an oil tank. This did not take long and was transmitted by e mail. Early the following year I was offered a contract by Indian Railways to attend at their various works for a duration of four weeks in order to advise on oil firing and other innovations to 'B' Class locomotives and commission the prototype 1001.

The system designed for Indian Railways was a direct copy of that fitted to FR locomotives. such as 'Linda'and 'Blanche'. However, Indian Railways in their wisdom departed from my design on three major elements:

1. The auxiliary manifold instead of being placed high up on the inside of the front sheet of the cab was set low down and on the outside. It could only be drained by climbing on to the top of the fuel tank and reaching down.

2. The atomising steam control valve was a 2:1 reducing valve with a spindle with a very fine thread. It was impossible to react quickly enough to variations in demand for steam. I must bear some responsibility for not having been more detailed in my specification, but felt sufficiently confident at the design in giving Indian Railways engineers more credit than they deserved for sound practice. This valve was a considerable impediment to a tolerable performance.

3. The blower was fed by the same pipe as that which supplied the atomising steam, and was starved of steam.

While the locomotive had been described as a 'new build', only the boiler could be described as new. The rest was a locomotive. in a very worn and decrepit condition. The dome was tilted to one side – reminiscent of an Emmet creation. But the most serious defects lay in the smokebox:

1. Most of the smoke box door ring was corroded away, leaving the door to close against a void. All steam locomotives are dependent upon a good smokebox door seal in order to support a partial vacuum in the smoke box. Without an effective vacuum, combustion will be seriously impaired, and oil firing is no exception. The only way in which a seal could be achieved was to pack the void with a thick coil of asbestos rope served with graphite grease, until there was sufficient padding for the door to close against.

   I found several holes in the smoke box barrel which were closed with bolts, but could not be sure I had found them all.

2. The blast pipe (like the dome) was tilted to one side, so the effectiveness of the blast was much reduced.

There was some sort of political embargo in operation, in that the Siliguri staff were reluctant to get involved with the 1001. GOC sent a team of four fitters and an engineer to carry out any commissioning work. They worked very hard one weekend to correct the mis-aligned blast pipe.

The locomotive. had a 'home made' displacement lubricator poorly fabricated from mild steel situated just above the atomising steam valve.

121

It leaked scalding water on to the atomising steam valve, (and my hand) making operation of the valve even more difficult and unpleasant.

The injectors were temperamental in the extreme, taking two minutes or more to get going. At one stage, it was necessary to shut the burner down because it appeared the injector had ceased working. With oil firing, injectors are used in a cyclic sequence: pressure rises to just below BP: injector 'on'; injector kept on until pressure falls to about 10 psi below BP; and so on. For this sequence to work, an injector must be got going within about 8 seconds. If not, firing quickly becomes problematic.

There was no compressed air at Siliguri Shed for lighting up 1001. The shed staff had resorted in true Indian Railways fashion to taking the air brake air from a diesel locomotive parked near by and piping it to 1001 in order to raise steam. This required the diesel engine to be run at near full revs for about two hours to achieve about 20 lb psi on 1001, at which point it was possible to change over to steam. For three days we had the use of a C-P towable diesel driven compressor procured through the good offices of Rajendra Baid of the Cindrella Hotel, which worked very well, before it had to be returned from whence it came. The next arrangement was to take steam from another steam locomotive parked near by, but manufacture of the connecting pipework took the best part of a day. However, it was effective and cut the steam raising time down considerably.

I was anxious to see how the locomotive performed in its present condition, and arranged for a trip up the line the following day I fondly imagined on that first day of steaming that we would have been 'off shed' by mid-day at the latest. But as with so many things in India, it was not to be.

As we prepared for our first trip up the line, I was told that technicians had been sent for from Tindharia to set the valves. I was a little taken aback, assuming they would have been set before my arrival.

We finally departed at 15.30, then to be held at the signal box for half an hour waiting to cross the broad-gauge line. And then there were the 'hangers on' – literally. Apart from the driver, fireman and me, there were two each side hanging on to the footsteps. They were probably the sand men and coal breaker. The run to Sukna is level, so there was never going to be a strong blast from the exhaust. But the weak blower made it impossible to compensate, and even running light engine, we inexorably lost pressure and water and stopped by the Mogmor tea plantation for a blow up. We arrived at Sukna about 17.00, and took water. I was expecting we would return to Siliguri, but to my delight we set off up the hill. I suppose Rangtong was the intended destination. On the hill, the driver opened up and I was able to establish steady state conditions with the locomotive starting to perform as it should. With the atomiser set, it was a relief to be able to keep my hands clear of the drip from the lubricator. Then disaster struck: on the first left-hand curve the locomotive derailed at a dipped joint. Running light engine, we had been travelling faster than a service train, and the dipped joint was too much.

It had been my hope that we could have steamed the locomotive. each day for 3-4 days in order to gain experience of its performance. Once warmed up, lighting up the next day would not have taken so long. But numerous obstacles intervened, and it was the following week before another trip took place.

By this time, the pipe for taking steam from an adjacent locomotive had been made and the alignment of the blast pipe corrected, although on further checking, it was set back from the centre line of the chimney by about half an inch. More fuel was delivered and the locomotive topped up. A separate steam supply for the blower had been taken from the driver's side injector.

The next trip was much like the first with little improvement apart from a more effective blower. When we got to Sukna, I expected we would continue up the hill as previously. But to my surprise and disappointment, we set off back to Siliguri having learned little.

More in desperation than necessity, I set about increasing the clearance of the oil gap of the burner. This required some shim steel, which I fondly imagined could be procured without too much difficulty. But it transpired shim was not readily available, and it meant a trip to the diesel depot and much rummaging around dusty stores before any was found. I made a shim washer and fitted it, but by now, it was Friday, so further steaming would have to wait until the following week – my last.

Cedric Lodge at Siliguri Shed on 16 February 2007 (Dave Priestley)

The locomotive was prepared on Monday for steaming on Tuesday, with the intention of making another trip up the line. As on the previous two occasions, as a result of Indian prevarication, it was mid afternoon before we departed, incurring the by now usual wait for a path across the broad-gauge. There was one significant improvement: the Shedmaster had conceded to my request that a flat wagon be attached for the multitude of hangers on to ride on, relieving the overcrowding of the locomotive. The performance was much the same despite the increased oil gap and improved blower. I was getting better at controlling the burner, but the fireman's side injector was more troublesome, sometimes taking several minutes to get it going. As before, we stopped for the predictable blow up. At Sukna, the injector gave up completely – or so I thought. Pressure had risen to 10 psi above BP, so I shut the burner off and expected arrangements would be made for a rescue locomotive. (I speculated at what pressure the safety valves would finally have lifted). But after ten minutes or more, the fireman jubilantly proclaimed he had got the injector going! I lit up, and with the blower full on, we restored the water level to respectability. Sadly, we did not go any further, and returned to Siliguri.

I had arranged with Subrata Nath, the recently appointed Director of the DHR, that we would make another trip the following day, but he did not show up until mid-day. When he instructed the staff to light up, he was told there was no more fuel. And so ended my dalliance with 1001.

There is no doubt in my mind that most of the ingredients were present for 1001 to work successfully on oil. But the project was seriously inhibited by irritating deficiencies ie; smoke box sealing, blower steam supply, atomising steam valve, leaking displacement lubricator, non availability of a compressor, and a reliable oil supply. With another three weeks at Siliguri, we could have got 1001 to perform as required.

Appendix 6 Baldwin Specification for B Class locos

(Courtesy De Golyer Library, Southern Methodist University, from MSS61 Baldwin Locomotive Works)

The source is pages 209 and 210 of a Baldwin Specification book held by De Golyer Library. The original was a standard 'pro forma' completed in typed capitals with additional notes, some typed and some handwritten. Some parts are very difficult to read and the following transcription has been made trying to maintain the style of the original by using italics for handwriting, with entries underlined as in the original. Some sections, for example those referring to front and back trucks, in which all entries were NONE or -------- have been omitted for brevity. The numbering of the supplementary notes at the end is as the original – that is 6, 7, 8 and 9 are missing.

333

THE BALDWIN LOCOMOTIVE WORKS AS SPEC'N. C-4258 AND Co's PRINTS. FOREIGN SPECIALITIES

Memorandum Spec'n 11189   Ry Co's Class

SPECIFICATION FOR LOCOMOTIVES 4 16 C 145 to 147   Gen as Ry.Co's Print 13-4577 *(see sup 2)*

FOR THE DARJEELING HIMALAYAN RY.   **Engine Drawing No** 24

M   INDIA   10-28-16

11 **Safety Valves** *(NOT SPEC)* TWO CONSOLIDATED 2" ENCASED

Gauge of Road 2' 0"   Play 20¾" between inside of tires   Safety valves NONE with relief lever, Set to 140 POUNDS

Wheelbase Rigid 5' 6"   Driving 5' 6"   Total 5' 6"   **Whistle** B.L.W. PLAIN

{ Actual ----   **Steam Gauge** ONE BRASS 6" Maker:- ASHCROFT *(NOT SPEC)*

Total wheelbase and tend. { Limited to ----   Steam gauge stand **BRASS**   Clock NONE

13 **Curves** 60 FEET RADIUS.   Grades 1 IN 30 OR 3 ⅓ PER CENT   *24* **Water Gauge** TWO B.L.W (IRONCLAD) **Cab Lamps** TO BURN OIL

{ TO TOP OF   **Gauge cocks** NONE   Surface Valve NONE

Limits, Height { STACK   8' 5¾"   Width ------   *Lowest reading of gauge glass 2¼" above highest part of crown*

Weight (approximate) in working order, Total 34,160 POUNDS   *NCW.12.4.17*

On F. tr'k ---- lbs., Drivers 34160 lbs, B. tr'k ---- lbs.   **Lubricator steam valve** TO HAVE separate dry pipe to dome

WEIGHT OF ENGINE EMPTY ABOUT 25,200 POUNDS   **Blower valve** CONNECTED TO RIGHT SIDE OF DOME

3 Tractive power 7750 lbs Ratio of adhesion 4.4 TANK EMPTY 4   **Blow off Cock** TWO B.L.W   POSITION RIGHT AND LEFT

Working St'm Pres. 140 lbs   Service SWITCHING   { **Cleaning plugs**

Fuel   SOFT COAL   { Firebox CORNERS AND FRONT

                                                                        { Crown   ONE BACK

2 **Boiler** STEEL   Thick ⅜"   Bill 7983   { Waist NONE

10 Boiler, Diam. 31" Plan STRAIGHT TOP   { Front tube sheet, ONE Fusible Plug FOUR LOCATED AS Ry Co's PRINT

                                { 160 lbs steam pres.   **ENAMELED STEEL COLLAR** RIVITED TO JACKET AROUND CLEANING HOLES

Boiler st'm pres. 140 lbs, Tested to { 187 lbs steam pres.   *W.O.W. 1.15.17*

                                                                        Pump   None   Pos. ----

Longitud. seams BUTT JOINTED, DOUBLE COVERING STRIPS,   Top chamb. ----   Bottom chamb. -----

QUADRUPLE RIVETED   Feed Cock -----   Feed pipes   . ----

11 **Dome**   diam 16"   Pos FORWARD OF FIREBOX   Checks ----   Check pipes   ----

2 **Firebox** COPPER   Combustion chamb. NONE   *(NOT SPEC)* **Injectors** TWO METROPOLITAN CLASS "H"   No. 3

Firebox length 36-7/16"   Width 35" Depth F 38¾" B 37-7/16"   Injectors, Pos. RIGHT AND LEFT   Steam pipes COPPER

Firebox, Tube ⅝" and ⅜"   Crown ⅜" Sides ⅜" Back ⅜"   Pipes, Check COPPER   Feed COPPER   Overflow IRON

Firebox opening Oval 10" x 12"   Check B.L.W.   Pos. RIGHT AND LEFT

Water space, Front 2"   Sides 2"   Back 2"   **Steam Heat** NONE

Water space, Frame SINGLE Riveted   *12* **Cab** STEEL CANOPY CRYSTAL PLATE GLASS CAB WINDOWS

Water space stays ⅞" DIAM. with 3/16" Holes 1¼" deep from O.S.   Cab Boards STEEL   Nosing ANGLE IRON

23 WATER SPACE STAYBOLTS OF COPPER   Running boards STEEL   Nosing ANGLE IRON

Radial stays 1 1/16" DIAMETER   Sling stays front NONE   Hand Rail IRON (SEE sup.*12*)

Button heads below crown on NONE   central rows   **Back Bumper** STEEL PLATE

Crown bars   NONE   Placed --- above crown

Crown bolts NONE   RAIL GUARDS TO BE PROVIDED FRONT AND BACK

{ Flues ----   **Pilot** NONE   Pushing shoes NONE

Tubes #12 and #14 W   W.G.Brass   **Front Bumper** STEEL PLATE   **Bumper plate** WITH

Tube ends STEEL THIMBLES IN FIRE BOX ENDS   **Steps** NONE

Number 65   Diam 1⅝"   Length 10' 1 3/16"

{ **Heating** { Flues ---- Combus. Chamb --- Total 316°   **F & B. Drawhead**   CENTER ABOVE RAIL 17"

{ **Surface** { Tubes 276° Fire Box 40° Fire Brick Tubes ---- **Coupler** RADIAL DRAWBAR FRONT AND REAR (see sup L3)

Grate area 8.9°   Ratio to heating surface 1:35.5   Signal fixtures NONE

Superheater ----

                                                                 SIX HEADLIGHT CHIMNEYS AND TWELVE WICKS

Fire Brick Arch NONE   **Headlight** ONE ROUND CASE 10" Reflector, TO BURN KEROSENE

Grates PLAIN BARS

Ash Pan Dampers FRONT AND BACK   Headlight, Nos. on sides NONE   H'dl't stop   NONE

                                                                 Lamp Shelf NONE   Nosing -------

Throttle BALANCE SLIDE POS. DOME

Dry pipe   COPPER   **Bell** NONE

Superheater NONE   *14* **Sand Box** TWO (see sup-L4) Body IRON PAINTED Pipes IRON

Smokebox extension NONE Cleaning hole & cap   NONE   Sander HAND, TO SAND FRONT OF FRONT WHEELS ONLY

Spark ejector or hopper NONE

Netting and deflect. plates NONE   Dome casing IRON PAINTED, HELMET SHAPED

Exhaust nozzles SINGLE   **Boiler Jacket** # 18 PLAN. STEEL Bands IRON Style 8 AND 14

Stack STRAIGHT   Dia 9"   Hole in S.B. dia 8⅞"   **Boiler Lagging**   MAGNESIA SECTIONAL

SMOKE STACK TO HAVE COPPER TOP FINISH   BACKHEAD OF BOILER TO BE LAGGED AND JACKETED

123

2 **Frames** OUTSIDE OF WHEELS, STEEL PLAT **Braces**

2 **Cylinders** Diam 11"   Stroke 14"
St'm ports, Length 8 ½"   Width 1"   Bridges 1 1/16"
Exh ports, Length 8 ½"   Width 2"   Valve lead 1/8"
Valve travel 3-5/32"   Steam lap ¾"   Exhaust clearance ---
Eccentric crank throw 4"   By-pass valve NONE
S.F Lubr. (NOT SPEC) MICHIGAN DOUBLE BULLS EYE 1 1/8 PINTS CAPACITY
17 Cyl'r heads CAST IRON: LUBR. PIPES OF COPPER O.S. OF JACKET
Cyl'r casings IRON PAINTED   Head covers NONE
St'm chest casings NONE   Covers NONE
St'm chest valves BALANCED OF BRONZE GUN METAL
~~Met~~ pkg {ISTON RODS/ VALVE STEMS   BRAIDED PACKING

18 **Pistons** CAST IRON RINGS SPRUNG INTO SOLID HEAD
Guides STEEL   Oil cups IN CROSSHEAD
Crossheads CAST STEEL (SEE SUP. 19)
Wrist Pins STEEL   Rod cups COMPANY'S STYLE

20 **Valve motion** WALSCHAERTS
{ Links, sliding blocks, pins, lifting links and eccentric rod jaws to
{ be well case hardened ~~or to have hardened bushings~~
All threads for bolts and nuts to be Whitworth Standard 11/20/16
~~All threads on bolts (except staybolts) to be United States Standard~~
All finished movable nuts of steel or of iron case hardened.

21 **Wearing Bearings** B.L.W.
DRIVING AXLE CRANKS OF CAST STEEL WITH HUB LINERS OF BRASS
Drivers, O.S. Diam 26" Cen.diam 22 ½", CENTERS OF CAST STEEL
DRIVING WHEELS TO HAVE HUB LINERS OF BRASS
22 **Tires** STEEL Held by SHRINKAGE AND SHOULDER AND TAP BOLTS
Tires, flanged, Pos ALL   Size 1 ¾" x 4 3/8"
Tires, Plain, Pos NONE   Size ------
Lateral play between driving boxes and wheel hubs = 1/32" TOTAL
Driving Boxes CAST ~~STEEL~~ BRONZE
Journals,   Diam 4 ¼"   Length 5"

Springs   Make B.L.W.

Front Engine Truck NONE   Wheeled ------
To swing ----- each side of cen., Cen. Pin liners ------
Journals, Diam. ---- Length ---- Boxes ----
Wheels, Diam. ---- Tread ---- Kind ----
Tires ----   Cen. Diam. ---- Size ----
Tires held by ----

Back Engine Truck NONE   Wheeled ----
To swing ---- each side of cen., Cen. Pin liners ----
Journals, Diam. ---- Length ---- Boxes ----
Wheels, Diam. ---- Tread ---- Kind ----
Tires ----   Cen. Diam ----   Size ----
Tires held by ----

22 **Brake,** position ON ALL WHEELS (NO TENDER)
Operating)} Engine, Sch HAND SCREW
Parts   )} Eng. Tr'k, sch -----
Foundation } Drivers, sch   B.L.W.
Parts   } Eng. Tr'k, sch ----
Air pump ----   Pos. -----
Train connections   -----
Train signal, sch. ---------
BRAKE TO BE OPERATED BY VERTICAL SHAFT AT BACK OF CAB
LE CHATELIER BRAKE, VALVE ON LINE OF LOWEST READING OF
GAUGE GLASS

124

TENDER, NONE

TANK OF HALF SADDLE TYPE TO BE PROVIDED AT FRONT END OF
BOILER, WITH A CAPACITY OF 150 IMPERIAL (180 U.S.) GALLONS
TANK BILL 4367   *(see sup 2)*

TANK TO BE PROVIDED AT FRONT END OF ENGINE BETWEEN THE
FRAMES, CAPACITY 230 IMPERIAL (276 U.S.) GALLONS
TANK BILL 4363   *(see sup 2)*

TEST COCKS TO BE PROVIDED FOR BOTTOM WATER TANK.

COAL BOX WITH A TOTAL CAPACITY OF 1680 POUNDS TO BE
PROVIDED IN FRONT OF CAB ON BOTH SIDES
                                  (see sup 2)

ONE BRASS OIL CAN, ONE COPPER TALLOW POT AND BUCKET
LIST OF TOOLS, SCH 32 TWO TRAVERSING JACKS WITH RATCHET HEAD
PHOTO SCH. 2   INSPECTION ~~B.L.W.~~ BY HUNT & CO { see sup 5
TO BE WEIGHED YES   MATERIAL SPEC'NS (SEE SUP.1)
CERTIFICATE OF BOILER TEST DUPLICATE
JUNIOR LAWS NONE   SHIPMENT { BEST FOREIGN STYLE
                              { ENGINE TO BE DISMANTLED

{ BRASS SHOP FINISH, No 3, EXCEPT INJECTOR CHECKS TO BE
{ ~~PAINTED~~ FINISHED   *RSmee*
MACHINE SHOP FINISH D 158 EXHIBITION FINISH   *1.21.17.*

B.L.W.   }
PATENTS   } NONE

PROPERTY PLATES NONE

~~PAINT LEAD COLOR WITHOUT STRIPING OR LETTERING OF ANY KIND~~
PAINTING :- ENGINE, STYLE ---- *(see ✱ below)* 1/30/17

LETTERING, NUMBERS AND STRIPING TO BE APPLIED BY RY. CO.
MARK NONE

1/20/17
OMIT FRONT NUMBER PLATE
" ROAD NOS. POS.   ROAD NOS.   NAME NONE
"   NONE

" LOCOMOTIVES TO BE SHIPPED WITH METAL WORK, ETC ALL
" PRIMED AND SURFACED FOR ITS PROTECTION AND
" PRESERVATION.
" SHIP WITH LOCOMOTIVES A SUFFICIENT SUPPLY OF NAPIER
" GREEN ENAMEL TO BE USED FOR PAINTING LOCOMOTIVES AFTER
" THEIR ERECTION. HEREWITH SMALL SAMPLE OF NAPIER GREEN
" ENAMEL. IN THE PREPARATION OF THIS ENAMEL, BE CAREFUL TO
" USE SUCH VARNISH AS WILL NOT CRAWL OR PERISH TOO SOON
" IN THE HOT CLIMATE.

1. THE PHYSICAL AND CHEMICAL PROPERTIES OF ALL MATERIALS USED IN THE LOCOMOTIVES TO BE IN ACCORDANCE WITH THE LATEST STANDARD OF THE AMERICAN SOCIETY FOR TESTING MATERIALS, AS FOLLOWS:

   STEEL
   - A10-14 STRUCTURAL STEEL FOR LOCOMOTIVES
   - A17-13 BLOOMS, BILLETS AND SLABS FOR CARBON STEEL FORGINGS
   - A19-14 QUENCHED AND TEMPERED CARBON STEEL AXLES, SHAFTS AND OTHER FORGINGS FOR LOCOMOTIVES AND CARS
   - A20-14 CARBON STEEL FORGINGS FOR LOCOMOTIVES
   - A21-14 CARBON STEEL CAR AND TENDER AXLES
   - A23-12 FORGED AND ROLLED, FORGED OR ROLLED SOLID CARBON STEEL WHEELS FOR ENGINE TRUCK, TENDER AND PASSENGER SERVICE
   - A26-14 STEEL TIRES
   - A27-14 STEEL CASTINGS, CLASS B MEDIUM GRADE (SPECIAL REQUIREMENTS FOR CASTINGS FOR RAILWAY ROLLING STOCK)
   - A28-13 LAP WELDED AND SEAMLESS STEEL BOILER TUBES, SAFE ENDS AND ARCH TUBES
   - A30-14 BOILER AND FIREBOX STEEL
   - A31-14 BOILER RIVET STEEL

   WROUGHT IRON
   - A38-12 LAP WELDED IRON BOILER TUBES
   - A39-14 STAYBOLT IRON
   - A40-13 ENGINE BOLT IRON
   - A41-13 REFINED WROUGHT IRON BARS

   CAST IRON
   - A45-14 LOCOMOTIVE CYLINDERS
   - A48-05 GRAY IRON CASTINGS

2. LOCOMOTIVES TO BE DESIGNED IN ACCORDANCE WITH RY. CO'S PRINTS OF THEIR CLASS "B" LOCOMOTIVES AS FOLLOWS :-

   - 13-4577   GENERAL DRAWING
   - 13-103   BOILER AND FIREBOX
   - 14-4613   FIREBOX TUBE PLATE
   - 13-104   SMOKE BOX
   - 12-2663   CYLINDERS
   - 12-4488   MOTION ARRANGEMENT
   - 12-4478   FRAME ARRANGEMENT
   - 12-4362   TANK
   - 16-1096   CAB AND COAL BUNKER
   - 16-896   PIPING ARRANGEMENT
   - L-25

3. TRACTIVE POWER IS CALCULATED AT 85 % OF WORKING PRESSURE

4. FURNISH ONE FULL SET OF TRACINGS PREPARED IN THE SAME MANNER AS TRACINGS BEING FURNISHED FOR 10-12 D 12 TO 406.
   ? FURNISH ALSO ONE SET OF PRINTS IN ADDITION TO THE TRACINGS ?

5. SEND TO R.W.HUNT & CO., COPIES OF ORDERS COVERING THE BOILER AND FIRE BOX STEEL, TUBES, AXLES AND TIRES.

10. CENTER OF BOILER ABOVE TOP OF RAIL 3' 11"
    BOILER TO HAVE FACTOR OF SAFETY OF 6 6
    FRONT TUBE SHEET TO BE 5/8" THICK.

11. DOME TO BE BUILT UP TYPE.
    SAFETY VALVES IN TOP OF DOME.

12. CAB TO CONSIST OF A ROOF OF SHEET METAL ON POSTS WITH A FLAT FRONT AND TWO SPECTACLES
    CAB NOT TO HAVE A BACK OR SIDES
    HAND RAIL TO BE PROVIDED AT BACK OF CAB TO PREVENT PEOPLE FROM BEING THROWN OFF.

13. DRAW GEAR TO CONSIST OF COUPLING BARS WITH PIN HOLES AT THE END WHICH ARE ATTACHED AT THE FRONT AND BACK OF ENGINE TO A SWINGING CURVED EQUALIZER WHICH PERMITS OF GREAT FREEDOM IN PULLING WAGONS AROUND CURVES OF 60 FEET RADIUS.

14. SMALL SAND BOXES TO BE PROVIDED BELOW THE SADDLE TANK.

15. BOILER LAGGING TO EXTEND TO FRONT TUBE SHEET WITH JACKET EXTENDED TO FRONT END OF SMOKE BOX.

16. STEAM CHEST VALVES TO HAVE RELIEF VALVES.

17. STEAM CHEST COVERS TO BE IN ONE PIECE, FITTED AT A SLANT.

18. PISTON RODS TO BE OF THE EXTENDED TYPE.

19. CROSSHEADS TO HAVE RENEWABLE BRONZE SLIDE BEARINGS.

20. MOTION WORK AND CRANKS TO BE OF THE OUTSIDE TYPE.
    REVERSE LEVER TO BE PROVIDED.

21. DRIVING AXLES, WHEELS AND TIRES TO BE MADE INTERCHANGEABLE WITH RY. CO'S CLASS 'B' LOCOMOTIVES.

22. ON ACCOUNT OF THE VERY LONG DOWNHILL RUNS ON WHICH THE BRAKES HAVE TO BE USED CONTINUOUSLY, IT IS NECESSARY TO APPLY THE TIRES WITH A VERY MUCH GREATER SHRINKAGE THAN IS CUSTOMARY FOR WHEELS OF THIS DIAMETER FOR ORDINARY SERVICE. GIVE THIS SPECIAL ATTENTION.

23. WATER SPACE AND RADIAL STAYS TO HAVE 11 THREADS PER INCH.
    Crown stays 1 1/16" diam. at ends and 3/4" diam. at centres. NOW 1·15·17

24. WATER GAUGES TO BE LOCATED AS SHOWN ON RY. CO'S DRAWINGS.

FOR HEREAFTER :- PISTON ROD EXTENSION GUIDE TO BE AS SHOWN ON RY. CO'S TRACING # 2 ★
DRIVING TIRE FLANGES TO BE AS SHOWN ON RY. CO'S TRACING
★#L   (TRACINGS FILED IN THIRD FLOOR VAULT)
RAISE QUESTION ABOUT MAKING ~~PROPER~~ COPPER STAY BOLTS 1" DIAMETER
PISTON ROD TO HAVE COLLARS AT TAPER FITS FOR PISTON AND CROSSHEAD
SEE LETTER DATED JULY 5TH 1917 FROM E.P.WILLIAMS, JR.,
FILE #17250   R.S.McC 10/29/17

For Hereafter. See Report of E.P. Williams Jr. June 9-1917. File 17250, referring to location of Injectors, Safety Chains on Bumper, draw bars, tank water Gauge cocks, cab posts, valve rod guids, springs, and steam chest relief valves.
Prints filed in Third Floor Vault.
R.S.McC. 8/10/1917.

Annotated cab view of a round-top firebox 'B' Class locomotive